Mathematical Ecology

Mathematical Ecology

edited by
Peter Allen (Brussels), Werner Ebeling (Berlin),
Manfred Peschel (Berlin), Peter Schuster (Vienna),
Yuri M. Svirezhev (Moscow)

Mathematical Ecology deals with mathematical models of evolution processes, in the biosphere as a unity of growth and structure-building, with spring-up of new species, their interaction and possible extinction, with the impacts of human activities on the environment and the corresponding consequences for the biosphere.

The whole field seems to be a certain amalgam of suitable system philosophy, in which real world phenomena are considered through an ecological pair of spectacles with the help of system methodology and mathematics.

All publications published herein are of interdisciplinary interest for ecology, biology, economy and technical engineering.

Applied Systems Ecology

Approach and Case Studies in Aquatic Ecology

by Friedrich Recknagel

With 61 Figures and 6 Tables

Akademie-Verlag Berlin 1989

Author

Dozent Dr. sc. Friedrich Recknagel
Technische Universität Dresden

ISBN 3-05-500475-2

Erschienen im Akademie-Verlag Berlin, Leipziger Straße 3—4, Berlin, DDR-1086

Lizenznummer: 202 · 100/552/89
Printed in the German Democratic Republic
Gesamtherstellung: VEB Druckerei „Thomas Müntzer“, Bad Langensalza, DDR-5820
LSV 1085
Bestellnummer: 763 813 2 (9110)
02500

Preface

In the context of this book, systems ecology will be understood as a connecting link between ecological phenomena, their scientifically founded formalization and their anthropogenous control in an objective manner. For the purpose of control, ecotechnologies are envisaged as being based on utilization of internal feedback mechanisms in ecosystems without weakening the ecosystem's integrity.

The general approach suggested refers to the following elements: real system, model, computer and decision maker. The following principles must be adhered to:

- System concepts have to take into account the determination of differentiated amplitudes of ecological process rates to a wider extent in order to gain better insights into the redundancy potential of ecosystems.
- Model concepts must be credible and correspond to control concepts.
- Computer concepts should permit comprehensive simulation of state and control spaces of ecosystems.
- Control concepts should belong to the off-line type, with an expert system being suggested as a future prototype.

Within the framework of case studies the approach will be applied to water types such as waste stabilization ponds, lakes and reservoirs. Great importance is attached to wide spectrum of water quality management problems and of waters under consideration, thus examining the validity of the approach, particularly of the model concepts proposed.

To overcome any constraints of simulation methods which may arise in uncertainties of ecosystems and models, new perspectives such as the use of AI-methods are pointed out. The example given for such methods is a deep expert system for water quality control of lakes and reservoirs.

The present results were generated through the use of an extensive software package, which attempts to document results by means of computer graphics, consistently. For this purpose, hardware made by Hewlett-Packard GmbH, has proven to be value.

This book was stimulated by Prof. Manfred Peschel. I would like to thank him for his suggestions and for his encouragement in the course of writing this book. The book is a result of ten years interdisciplinary research by hydrobiologists, limnologists, physicists and system engineers at the Hydrobiological Laboratory of the Department of Water Sciences at the Dresden University of Technology. I want to thank my colleagues who were involved in this research for their co-operation. I feel especially obliged to Prof. Dietrich Uhlmann for his encouragement and for giving me valuable support for modeling in waste stabilization ponds. I would also like to thank Dr. Jürgen

Benndorf for his close co-operation in the development of the model SALMO. His causal research of lake ecosystems was of enormous value.

I am grateful to Prof. Grau, University of Chemistry and Technology Prague, Dr. Straskraba, Czechoslovac Academy of Sciences, Dr. Klapper, Dr. Ackermann, Mr. Pütz and Mr. Beuschold, Water Management Authority of the GDR, for valuable discussions, as well as to Dr. Horn, Dr. Paul, Dr. Koschel, Mr. Kneschke and Mr. Kruspe for making data of the waters which were under consideration available. I am also deeply indebted to Miss Mairead Scanlon for improving the English text, Mrs. Dietz and Mrs. Viehweg for some graphics and Mrs. Proft for typing the text.

I am grateful to the editors from Akademie-Verlag Berlin for their efforts to ensure a quick publication.

Dresden, December 1988 F. Recknagel

Contents

1. Introduction

In the process of decision making, the decision maker is always confronted with an investment threshold and an innovation threshold which are interdependent to a certain extent. Ideally, the investment threshold value should not remain below a certain level whereas the innovation threshold value can be reduced to a minimum by improving methods and tools in decision making.

The heuristic method as an inherent part of each decision making process is intuitively applied in order to utilize the decision maker's learning capacity. In Fig. 1, the principle of the heuristic method is illustrated. Irrespective of whether the decision is a failure or a success, the heuristic method results nevertheless in an increase in knowledge, be it through „good" or „bad" experience. However, the consequences of each decision are quite different. The consequences of successful decision making are characterized by time and investment efficiency along with innovative measures. In the case of a wrong decision however, the consequences may be as follows:

— losses in time and investment which in turn hinder the implementation of any innovative measures to solve the basic problems
— secondary and tertiary problems arising from wrong decisions which can be irreversible in extreme cases.

If the response behaviour of a real system is used to assess the effects of a decision, then it is said to be a direct application of the heuristic method. The quality of the decision is only tested on the real system which has to bear any consequences. In the past there was often little alternative to the use of heuristic principles on real systems, but the practice continues to be widespread even as alternatives have become more available.

On the whole, the applied systems analysis deals mainly with an indirect application of the heuristic method for decision making or policy planning. On this case, the real system is firstly replaced by an appropriate „surrogate" — usually a mathematical model — and in order to assess the decision or policy, it examines the response behaviour of the „surrogate" which has been made accessible by computer simulation. With this approach, the real system can be protected from many of the adverse consequences of „wrong" decisions which would have been a product of trial and error learning. Consequently, applied systems analysis can be defined as an attempt to objectify the decision making process with the aim of optimizing the investment threshold value and of minimizing the innovation threshold value of the decision maker by means of modeling, simulation and artificial intelligence. In the context of this book, applied systems ecology is introduced as a synonym for objectifying the control of ecosystems by including the applied systems analysis.

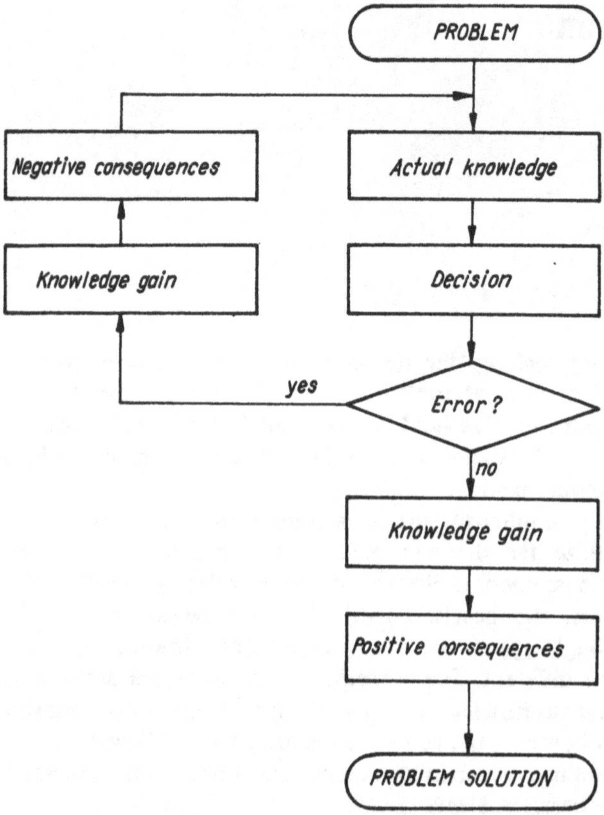

Fig. 1: Heuristic approach to decision making

The task of objectifying decision making for aquatic ecosystems has proved to be especially complicated. Firstly, the management problem involves various criteria as there are, at the very least, two targets to be reached:

— maximization of water quality
— minimization of costs.

These targets should not be considered separately but rather as being in conflict with one another. Therefore during decision making must be made a compromise. At the same time, the decision or policies should meet all the requirements of ecotechnologies (Uhlmann, 1983; Straskraba, 1985) by adhering to the main features of an aquatic ecosystem characterized by high complexity, nonlinear dynamics, spatial and temporal heterogenities and natural stochasticity. An adequate understanding of the underlying processes of ecology (which can also be gained by applied systems analysis) is necessary.

Generally, aquatic ecosystems are subject to many pollutants arising mainly from using the drainage area for various purposes and also from using the waters, e.g. for intense fishing or for recreational purposes. As a result of these pollutants, the quality of the water is considerably impaired and the desired uses of the waters, in turn, are at risk. In Fig. 2, the pollution/using relations between drainage area, water and potential users of standing waters are roughly represented.

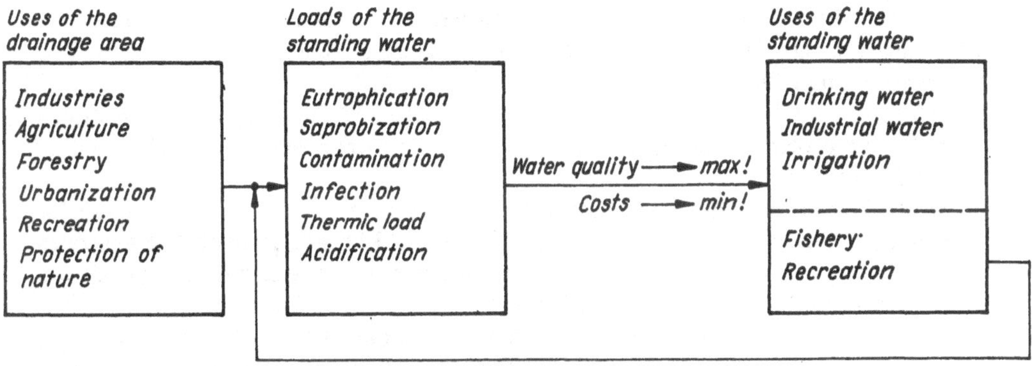

Fig. 2: Pollution/using relations of standing waters

Among the types of loads of standing waters, especially of lakes and reservoirs, the eutrophication is a special category. According to UHLMANN (1979a), eutrophication is described as a process whereby the waters are polluted by vegetable nutrients. This presents a global problem and is rapidly on the increase (VOLLENWEIDER, 1981). Eutrophication can be characterized by the following aspects:

- It has point and nonpoint sources of nutrients, so equally, external and internal control measures should be considered.
- It occurs in combination with other types of pollution (e.g. contamination due to application of biocides in agriculture and forestry).
- It causes other types of pollution (e.g. saprobization and contamination due to increased organic matter formation by aquatic primary producers, particularly phytoplankton).
- Other types of loads are used as catalyzers (e.g. thermal load).

In Chapter 2 of this book a general approach to applied systems ecology is put forward and consists of the following elements: real system, model, computer and decision maker. The elements are connected by the relations: modeling, simulation, decision making and control in a closed loop manner. To make the approach relevant for application in aquatic ecology, adequate concepts of systems, models, computers and control strategies are outlined.

The system concepts make allowance for the feature of aquatic ecosystems in composition, environment and structure. General system properties such as autonomy, coherence and resilience are considered in connection with the essential ecosystem processes: growth, self-reproduction and evolution.

As model concepts state description and input/output description are distinguished. Both concepts are considered with respect to the simulation of ecosystem's transient behaviour or steady state behaviour, respectively, and their relevance for ecosystems is discussed. A special note is given in the case of model validation to increase credibility of model applications.

The computer concepts serve for opening up such digital simulation techniques which permit the comprehensive analysis of state space or control space of ecosystems, respectively. With regard to the control concepts, perspectives of real-time simulation of ecosystems are discussed.

The control concepts are subdivided into off-line and on-line control. In spite of the fact that both control concepts are characterized by different temporal resolving power, demands are defined for the underlying model and computer concepts as well as for the measuring of ecosystems.

In Chapters 3 to 7 the general approach to applied systems ecology is illustrated by results and experiences in modeling and model-based water quality control of ponds, lakes and reservoirs.

Chapter 3 deals with the development and validation of an input/output model for waste stabilization ponds. Based on this model Chapter 4 illustrates an off-line approach for the optimum control of substrate removal performance of waste stabilization ponds and outlines a case study for a real pond system.

Chapter 5 deals primarily with the state-determined model SALMO. The model was developed to simulate reservoirs and lakes. The model is documented by state variables and process rates and is thoroughly validated. Within validation, a comprehensive sensitivity analysis is made and for several very different standing waters, measured and simulated trajectories are directly compared.

In Chapter 6, on the basis of the model SALMO, an off-line approach is developed for the optimum control of the water quality of reservoirs and lakes. The methods of experts estimates, scenario analysis, stability analysis, dynamic optimization of costs and risk analysis are applied in an integrated form. Scenario analysis is based on simulation of response behaviour of water systems in the „what if"-mode. It objectifies the selection of rational control policies from a set of feasible alternatives. For stability analysis, in the sense of catastrophe theory, the topology of simulated isocline surfaces in relevant control spaces is investigated. In this way, stable regions in control spaces are localized. Dynamic cost optimization is based on a scalar optimization with state space constraints. It is suited for the generation of optimum control trajectories where at the same time maintaining obligatory water quality standards. To assess the reliability of model-based decisions in terms of probability, risks analysis are executed using the Monte-Carlo-simulation.

In Chapter 7, a case study on the model-based eutrophication control of a multi-purpose reservoir is illustrated.

In Chapter 8, the limits of model-based decision making are discussed by the hand of uncertainties and perspectives are pointed out which open up the so-called knowledge engineering. A deep expert system for water quality control of lakes and reservoirs is proposed which combines simulation and AI-methods in an integrated manner. In future, this can be regarded as a prototype of an off-line control concept for aquatic ecosystems.

2. Modeling, Simulation and Control of Ecosystems

The theory of modeling and simulation suggested by ZEIGLER (1976) was taken as a basis to formulate an approach to applied systems ecology. As shown in Fig. 3 it consists of the elements: real system, model, computer and decision maker, which are related through modeling, simulation, decision making and control. The approach occurs in a spiral mode contributing to a more and more causally determined ecosystem control in the sense of ecotechnology. With regard to aquatic ecosystems it is attempted subsequently to define relevant concepts of the elements and relations of the approach.

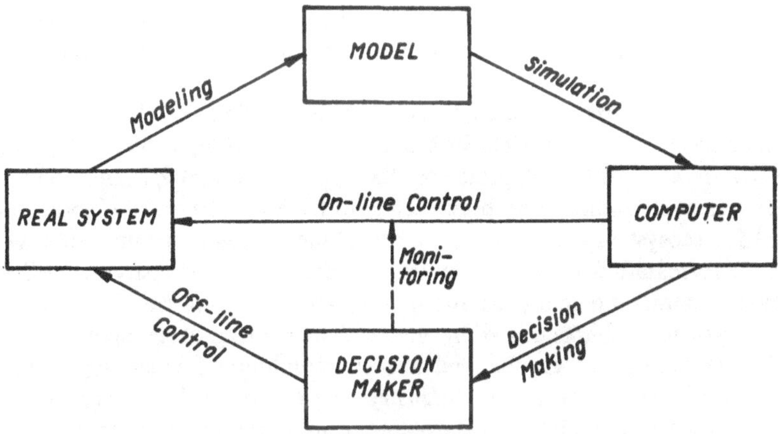

Fig. 3: Formal approach to applied systems ecology

2.1. System concepts

The whole is more than the sum of its parts.
Aristoteles

Nothing is permanent except change.
Heraklit

A system is generally characterized by its composition, environment and structure (BUNGE, 1979). The *composition* denotes the set of system components, the *environment* denotes the set of environment components which influence the system components. The definition of the composition and environment in turn implies the marking of

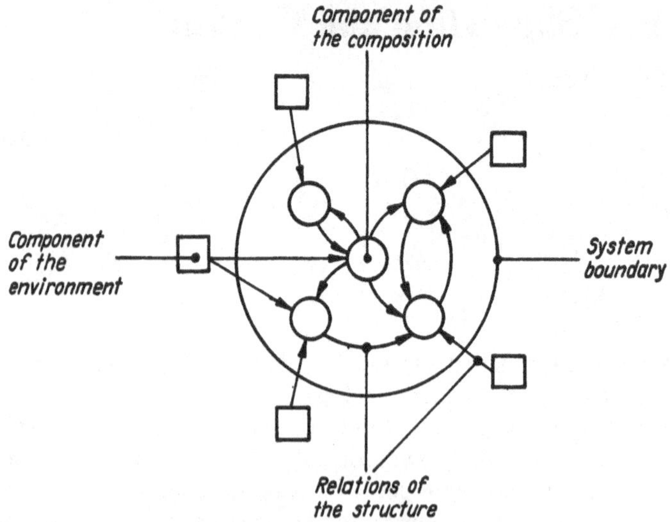

Fig. 4: General system concept

the system boundary. The *structure* denotes the set of relations between composition and environment as well as within composition. The general system concept defined is represented in Fig. 4.

In the case of an aquatic ecosystem, this concept is specified as follows. The composition of an aquatic ecosystem always includes biotic and abiotic components, in this connection, the diversity of composition depends on the type of ecosystem. The composition of fish ponds and waste stabilization ponds tends to have predominantly biotic components specialized for ecosystem's function and a multitude of abiotic components. Usually, in lake ecosystems manifold biotic components (a high species diversity) as well as abiotic components (organic and inorganic substances) are representative. The environment of an aquatic system is characterized by material and energetic components, which are usually characterized by geographic and seasonal conditions (meteorology) and by types of utilization of the drainage area (industry, agriculture, urbanization, protection of nature). They may have inhibitory and excitatory influences on the system components.

By means of this general definition, fundamental macroscopical system qualities have already been explained. Each system is of an *autonomy* which is maintained by the external interactions between composition and environment. Each system has a *coherency* (degree of integration) depending on structure and composition. If the internal relations are highly marked compared to the external relations, and their influence is excitatory, the coherency is considerable and the system is stable. If the internal relations are weakly marked as compared to the external relations and if they have an inhibitory influence, the coherency is slight and the system is fragile. For each system of a definite environment there is an optimum composition, a threshold composition and a maximum composition. The optimum composition gives rise to a maximization of the coherency. When the composition sinks below the lowest threshold, the system ceases to exist and when the maximum composition is exceeded, the system collapses. Each system is characterized by a certain *resilience* within a definite range of tolerance. It is characterized by the redundancy of the system in the composition and structure. With increasing redundancy the resilience of the system increases, compared to temporal changes of environment. Autonomy, coherency and resilience must not be considered separately since they are brought about mutually.

Because of the spatial unity of the water body of aquatic ecosystems a distinct autonomy is present which is at risk in extreme meteorological conditions (floods, long dryness) by external relations from the system environment. Aquatic ecosystems are hierarchically structured, because subsystems are formed by different population of biotic components. The relative autonomy and coherency of the subsystems competes with the total systems coherency. Regarding the resilience of aquatic ecosystems it is observed that dominancy and diversity of microorganisms within the composition renders high redundancy to the system and thus, the resilience is increased (e.g. waste stabilization ponds) whereas dominancy of more highly developed organisms gives a comparatively low redundancy to the system and thus low resilience (e.g. fish ponds). The reason for this are the different process rates of growth and self-reproduction of the species being dominant in the system.

Each system is subject to an irreversible *development* set off by continuous, extensively indefinite changes in the environment. The transfer of changes of the environment to composition and structure takes place by continuous energy transfers which from the thermodynamical view result in the dynamic formation of dissipative patterns in the system (Glansdorff and Prigogine, 1971). Dissipative patterns are spatially and temporally asymmetric inhomogeneous structures destined to approximate a stable equilibrium.

Aquatic ecosystems are predestined to form dissipative patterns under the influence of changes of the environment. The continuous inflow of intense energy by light and chemical energy activates processes in organisms and between them thus enabling their growth, self-reproduction and the compensation of inevitable side effects of the dissipation. As a result of these processes heat and low grade energy are carried off. Thus, the interplay of all system components ensures the ecosystem's steady-state. This means that the system must have a sufficiently diverse species which interact to such an extent that the energy dissipation is minimized. The arising dissipative patterns are maintained as long as continuously occuring changes in the environmental conditions cause bifurcations resulting in inhomogeneous repatterning of the system. The sequential repatterning of an ecosystem leads to the *evolution*. The *self-reproduction* of the organisms is to be regarded as internal basic process. Conrad (1983) characterizes the evolution under the following aspects:

1. The variability of the organisms caused by errors in selfreproduction
2. The natural selection of organisms which made possible through selective control of self-reproduction of the individual species by modified process rates
3. Strong principles of inheritance on the self-reproduction.

Consequently, the development of ecosystems can causally be explained from the microscopial view, i.e. from viewing the cell as the smallest unit of the self-reproduction. Even at cell level information processing takes place, thus enabling the system to control the process rates of self-reproduction of organisms differently and consequently, to minimize the energy dissipation. In this context a promising approach is opened by the concept of „hypercycles“ elaborated by Eigen (1971) and pursued by Eigen and Schuster (1979) and Peschel and Mende (1986).

To investigate the system behaviour in time the system dynamics concept established by

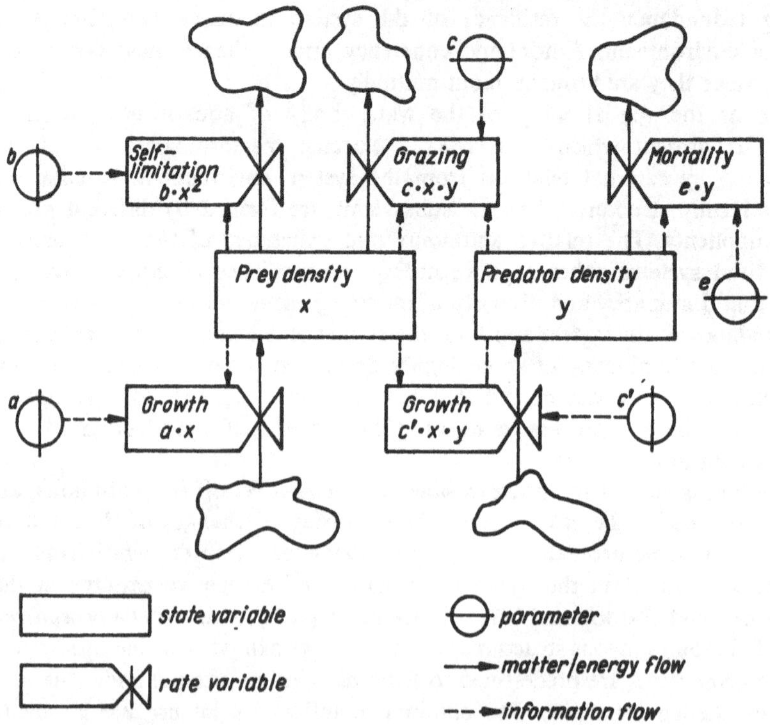

Fig. 5: Forrester-diagram of Volterra's predator-prey system

FORRESTER (1961, 1971) provides an efficient approach. Explicitly, this formalism allows to consider the process rates and the system components. The formalism replaces the process rates by rate variables and the system components by state variables. The environment components are considered as driving variables, constraints due to the system boundary are approximated by parameters. The interactions are separated in matter/ energy flow and information flow.

The Forrester diagram in Fig. 5 clearly illustrates the causal determination of the predator-prey system according to VOLTERRA (1926). The corresponding equations and simulation results are included in Section 2.2.2.

The aim of system dynamics was to facilitate the access to computer simulation. The instructions of the ad hoc defined simulation language DYNAMO (PUGH, 1971) refer to the symbols of system dynamics. As proposed by BURNS (1977) and RECKNAGEL (1980) and RECKNAGEL and BENNDORF (1978) based on the Forrester diagram and the related causal matrix and by using a dimension analysis of well-defined system quantities, computer aided identification of state and rate equations of the system considered can be performed.

The theory of living systems substantiated by MILLER (1978) gives access to the formalization of energy dissipation control in living systems on the basis of internal information processing. Miller differentiates between 7 system levels. These are: cell, organ, organism, group, organization, society, supranational system. To describe the processes of material processing, energy processing and information processing, he introduces 19 subsystems and he proves the evidence of each system level. He starts from the basic assumptions that each information is bound to material/energetic information carriers, and that in living systems 3 types of control always occur simultaneously:

1. Control of energy supply of the system composition by the environment,
2. Enzymatic control of energy transfers between environment and biotic system components and within the biotic composition,
3. Genetic control of the total development of the biotic components.

To formalize an ecosystem, HAUG (1983) applied the theory of living systems. For this purpose, he summarized Miller's 19 subsystems and grouped them into 8 aggregated subsystems (see Fig. 6). The resulting system concept in Fig. 6 requires a completely new approach to the ecosystem, because all the components must be interpreted and quantified from the functional view of the 8 subsystems. Those processes having been brought about by the system components responsible for transport, storage, conversion and reproduction of matter, energy and information come stronger to the force. In this connection, special attention is to be payed on the investigation of patterns of enzymes of dominating species, in order to reveal their strategies in differentiated control of process rates in the system.

For the identification of systems an experimental frame has to be defined for which two fundamental principles established by KALMAN (1979) must be taken into consideration: observability (measurability) and controllability of systems. *Observability* is based upon the assumption that each system state can be determined by the information included in the output of system. A prerequisite is to estimate nonobservable system quantities, as these are indirectly revealed by their effects on measurable system quantities or indicator variables. *Controllability* requires that each state of a system is determined by the system's input. This is necessary to solve optimum control problems. Both principles represent considerable idealizations which can only be approximated by appropriate reductions of the system. Elimination of nonmeasurable system quantities can bring about observability

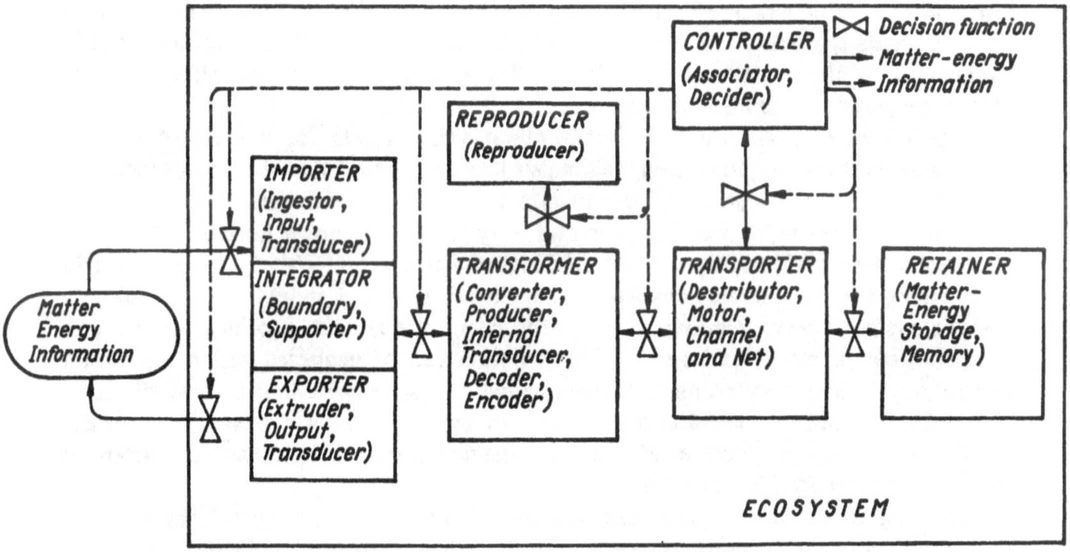

Fig. 6: Ecosystem concept based on Miller's living systems theory. After HAUG (1983)

of a system, elimination of uncertain system quantities can bring about controllability of a system. The combination of the two reduction types results in a canonical system which is observable as well as controllable.

To consciously pursue both principles is especially of great significance for those systems, on which there is only some a priori information and which must be inductively described by experimental investigations as it is relevant for ecosystems.

2.2. Model concepts

A model is a summary of experimental data; repeating an experiment on the model should yield exactly the same data as was assumed in constructing the model.

R. E. Kalman

In the construction and application of models three phases are typically:

1. Analysis, where for given input quantities of a system the resulting output quantities are identified.
2. Synthesis, where for given input and output quantities of a system the pertinent model is determined.
3. Control, where for the given model and the given output quantities of a system the input quantities are determined.

Analysis and synthesis characterize the modeling process. The control phase takes place within the simulation process. Analysis represents a direct problem because for known input quantities of a system a definite set of output quantities always exists. Synthesis and control represent inverse problems, for which an infinite number of solutions exists. Credibility of a model depends on reliability of the solution of inverse problems. The move from synthesis to control, i.e. from modeling to simulation, requires therefore a careful examination of the model's validity which is based on three criteria according to LEVINS (1966): generality, realism and precision.

A model gains *generality* if it brings about valid results for a class of systems. Depending on the well-defined composition, structure and environment of the system class, the model within this cut-out of the manifold pattern of conditions of a real system is examined in order to see if it generates a systems adequate behaviour. *Realism* of a model is determined to a large extent by the degree of abstraction of the system in the model which should be in a direct proportion to reality. The model should always reflect the essential system aspects. *Precision* refers to the extent of quantitative coincidence between the behaviour of model and system. This criterion can be neglected within the off-line control for the long-term planning where qualitative system behaviour is primarily taken as a basis. For on-line control in a short-term or real-time status, precision is of utmost significance. Here, the model is calibrated for the actual measuring values of a system by means of appropriate filter methods.

According to the afore mentioned validity criteria we can distinguish between the following four model generations (see Fig. 7): there are descriptive, predictive, prescriptive and calibrated models.

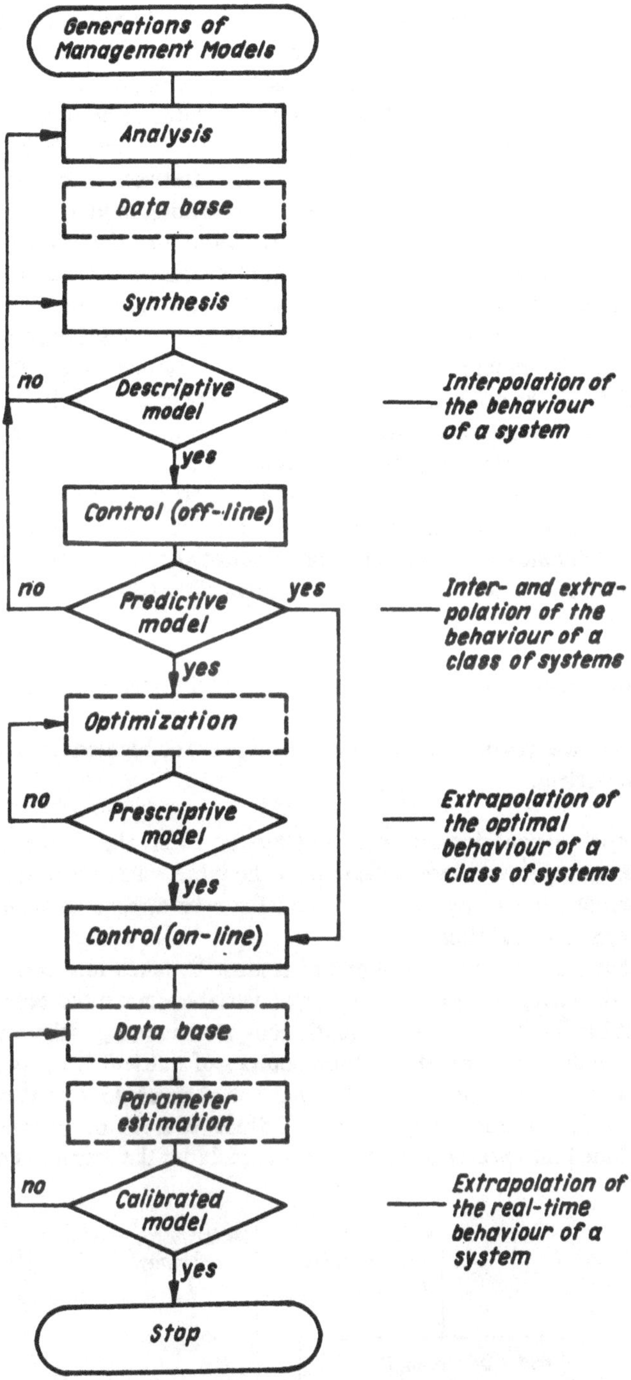

Fig. 7: Generations of management models

Descriptive models generate relatively valid output quantities for known input quantities of a specific real system. Because of their composition and structure, they prove a certain realism, however, their generality can not be proved. Thus, the inverse problem of the control phase remains unresolved. Descriptive models are scarcely suitable to be used as decision making tools, however, they can be applied as research tools to test scientific hypotheses. In case of *predictive models* it is presumed that they are descriptive for several real systems, i.e. for a broadest possible system class without calibrating model's composition, structure and parameters externally. Under those preconditions, the model acquires a certain generality and the inverse problem of the control phase can be regarded as solved within the specified system class. Predictive models are well suited to support decision making in medium-term and long-term planning, where its „sphere of competence" is always demarcated by the boundaries of the considered system class. *Prescriptive models* must be predictive and the optimum input quantities in relation to well-defined goal functions are generated for a real system. Solving the inverse problem, a higher degree of credibility is obtained by including appropriate optimization methods into the control phase. Within the frame of decision making, prescriptive models can help solve optimum control problems in the long-term, medium-term but also in the short-term status. *Calibrated models* will be obtained when predictive or prescriptive models are embedded in appropriate filter algorithms, e. g. into the extended Kalman filter (BECK, 1983). The filter carries out two functions:

1. Estimates of the model's state variables and parameters are calibrated by actual measured data of the system.
2. Measuring errors and influences of exogenous disturbances are eliminated from the actual measured data of the system.

The precision of a model is considerably increased by the filter algorithms, however, at the same time it is necessary to measure continuously the input/output behaviour of the system. Under these preconditions, the calibrated model can be applied to solve control problems in the short-term and real-time status.

In the modeling process, the system quantities which must be taken into consideration are represented in Fig. 8. Vector ***E*** represents the input variables measured continuously or periodically. Measurable input variables of aquatic ecosystems are e.g. substrate inflow concentrations of waste stabilization ponds or point sources of nutrient loads of lakes or reservoirs. Vector ***S*** represents the nonmeasurable input variables assuming the character of disturbance variables in the modeling process. Diffuse sources of nutrient loads of waters should be included into vector ***S***. State vector ***Z*** includes the essential conserved

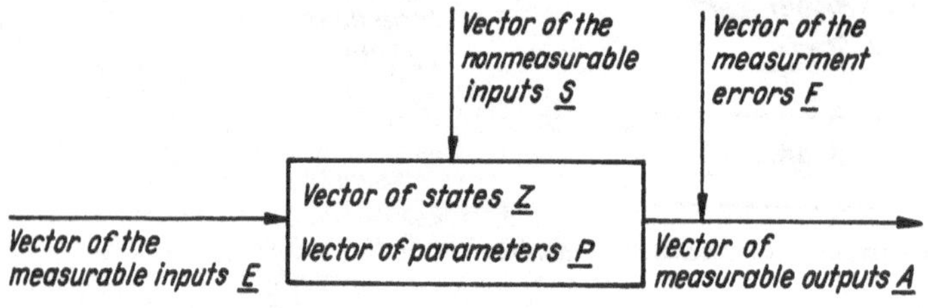

Fig. 8: Formal system representation

system quantities of physical, chemical and biological nature such as dissolved oxygen or the biomass of dominant species. As a rule, the measurable output variables $\boldsymbol{A}$ refer to the system's state quantities.

Parameter vector $\boldsymbol{P}$ represents essential coefficients to quantitatively determine the state or output quantities. It is a question of statistical coefficients or of well established coefficients, e.g. half saturation constants of growth kinetics of organisms. Each measurement is subject to random measuring errors and systematic measuring errors resulting in a more or less considerable falsification of the input and output quantities measured. They are summarized in vector $\boldsymbol{F}$. The system quantities characterized in this manner allow two alternative approaches of modeling to be used: input/output determination and state determination of systems.

2.2.1. Input/Output determination

Input/output determination can be applied to model the steady state behaviour of systems. Under these circumstances, the dependence of the output quantities $\boldsymbol{A}$ can thus be determined from the input quantities $\boldsymbol{E}$ without taking the dynamics and interactions of endogenous system quantities into consideration:

$$\boldsymbol{A} = \boldsymbol{f}(\boldsymbol{E}, \boldsymbol{P}) .$$

Vector $\boldsymbol{f}$ represents the set of transfer functions which are always of a correlative type, vector $\boldsymbol{P}$ represents the set of estimated coefficients. To determine the transfer functions, comprehensive statistical standard software for different computer systems are at disposal. In this way, the result of the synthesis phase of modeling is less dependent on the reliability of synthesis methods than on the precision and relevancy of the data base for the system or for the system class gained in the analysis phase. When there is no well founded data base which can be taken as a basis for the synthesis phase, the subsequent control phase should be subject to appropriate methods of error and risk analysis. They can be based upon the Taylor series expansion (DITLEVSEN, 1981) or on Monte-Carlo methods (MONTGOMERY, LEE and RECKHOW, 1980).

In the water quality management of lakes and reservoirs, steadystate models according to the phosphorus loading concept by Vollenweider (VOLLENWEIDER and DILLON, 1974; VOLLENWEIDER, 1976) are extensively applied and modified (e.g. RECKHOW, 1979; BENNDORF, 1979a; RECKHOW and CHAPRA, 1983). They enable us to estimate the trophic state and the chlorophyll-a of waters by calculating the phosphorus concentration P mg/l in the water as a function of areal phosphorus loading L mg/m^2 · year, the mean depth $\bar{z}$ m of water, the mean hydraulic residence time $\bar{t}$ year and the phosphorus removal rate by sedimentation $S(\bar{z}, \bar{t})$:

$$P = L/(\bar{z}/\bar{t} + S(\bar{z}, \bar{t})) .$$

An estimation of the new trophic state of standing waters, following a control measure, is provided by a phosphorus criteria plot like that shown in Fig. 9. The critical phosphorus loading L_c in mg/m^2 · year is calculated as follows:

$$L_c = 10(\bar{z}/\bar{t})\,(\bar{t} + \sqrt{\bar{t}}) .$$

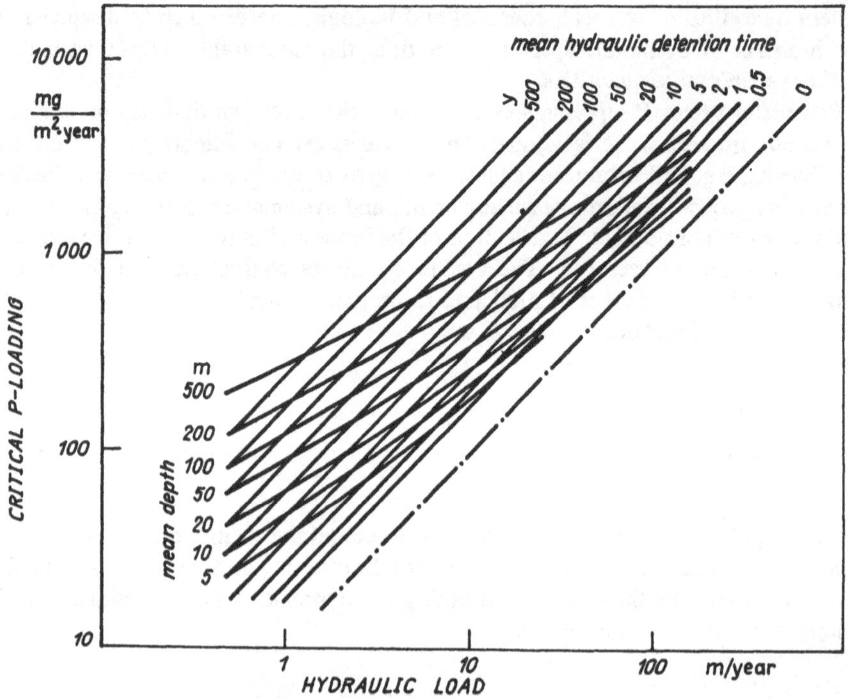

Fig. 9: Vollenweider's phosphorus loading criteria plot. After RECKHOW and CHAPRA (1983)

The reliability of predictions generated by those models will always depend on keeping to the limits of climatic, morphometric, hydrological and chemical conditions as well as maintaining the steady state condition which is demarcated by analyzed data base. In this context, CHAPRA (1979) discusses the influence of subjective perception errors in the analysis phase — caused by measuring errors — and the seasonal variability of waters on the data base. As generally the data base is obtained by measurements of different waters, CHAPRA (1979) emphasizes a certain incompatibility of the data base and the „originality" of the individual waters.

In Fig. 10, the result of an error estimation in the use of the phosphorus loading model applied to Liberty Lake (USA) as proposed by RECKHOW and CHAPRA (1983) is illustrated.

Such objective restrictions of correlative models can be overcome by an efficient measuring system or by means of modern parameter estimation methods. By means of the Kalman filter, for a given model the parameters and state variables are estimated using the measured input and output quantities.

The reliability of the data base for the synthesis of correlative models can be increased a priori, in case the input/output behaviour of the system can be simulated by an adequate laboratory simulator within a representative spectrum of conditions. In Chapter 3, using the example of the substrate removal model for waste stabilization ponds, it is demonstrated, that the statistical planning of experiments including an ecological laboratory simulator provides the data base for synthesizing an input/output model of predictive validity. Summing up, it is assessed that manifold applicabilities for the off-line and on-line control

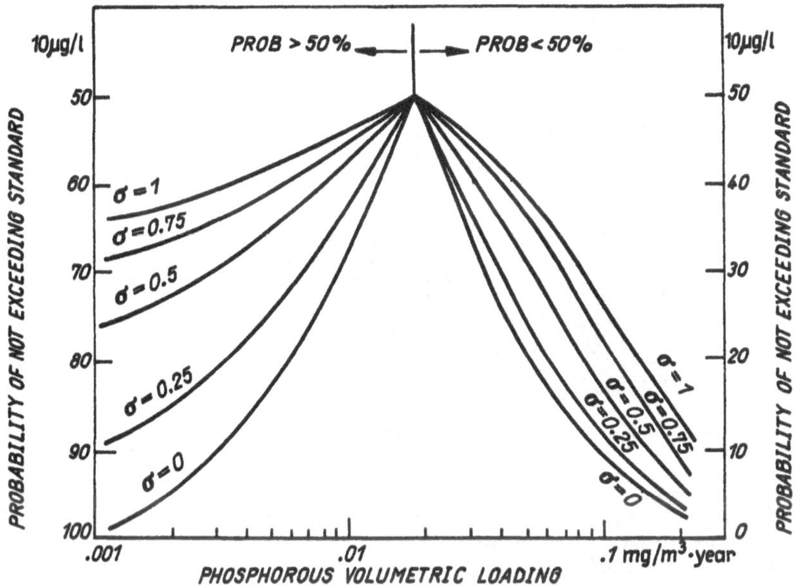

Fig. 10: Error estimation in the use of a steady-state phosphorus model for Liberty Lake (USA). After MAHEMAH and BHAGAT (1983)

of aquatic ecosystems are provided by predictive input/output models. The simplicity of their mathematical representation is a noteworthy advantage, thus, their implementation in microprocessors is feasible and consequently, there is access to a broad application of microprocessors.

2.2.2. State determination

The state determination is suited for modeling the dynamic or transition behaviour of systems. In accordance with Fig. 8, the formal description of the state determination is:

$$\dot{Z} = f(Z, E, P, t),$$

$$A = g(Z, E, P, t).$$

In these equations, the state vector Z, the vector E of the input quantities, the vector A of the output quantities, the parameter vector P and the time t are taken into consideration.

In the case that $E = 0$, an autonomous model is obtained. A stochastic or deterministic model is obtained depending on whether in addition the stochastic disturbance variables S and measuring errors F are considered or disregarded. In the stochastic model, the parameters are considered in a fuzzy manner by probabilistic distributions, in the deterministic model they are precisely estimated. The development of stochastic models is illustrated by FINNEY, BOWLES and WINDHAM (1982).

The predator-prey model suggested by VOLTERRA (1926) is a well-known autonomous deterministic model in ecology. Volterra considered the interactions between a prey species,

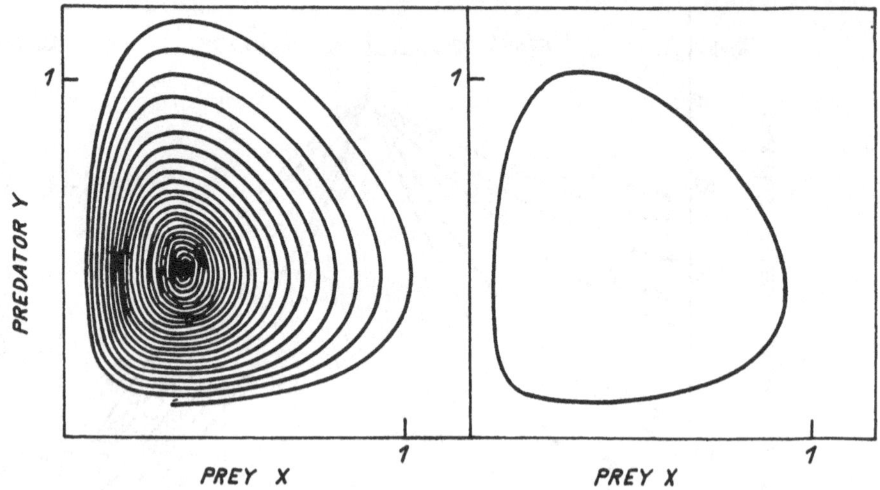

Fig. 11: State space representation of the solution of Volterra's predator-prey model
a) Consideration of self-limitation of prey
b) Elimination of self-limitation of prey

density x, and its predator, density y, where time-invariant rate coefficients were assumed:

$$\frac{dx}{dt} = ax - bx^2 - cxy\,,$$

$$\frac{dy}{dt} = -ey + c'xy\,.$$

In Fig. 11a, state space representation of model solution is given by $x_0 = 0.3$, $y_0 = 0.1$, $a = 7$, $b = 0.5$, $c = c' = 15$, $e = 5$. The Fig. 11b shows the solution of the modified model with $b = 0$ for eliminating the self-limitation of the prey and $a = 6$. In the following chapters, exclusively nonautonomous deterministic models will be considered.

Generally, the transition functions $\boldsymbol{f}$ and $\boldsymbol{g}$ are strongly nonlinear and are of causal nature. To determine the transient functions, a good knowledge of systems and a profound data base are necessary. The data base requires a high level of measuring techniques as it has to record the systems dynamic behaviour. In addition to the precise continuous or periodical recordings of input and output quantities by field investigations to further qualify state determined models, increasingly, rate variables must be quantified. The rate variables represent the internal control processes regulating the matter and energy flow in a system (see Section 2.1). Thus, they take a key position in balancing state quantities of a model. Generally, measurements of rate variables for optimum and pessimum conditions can be made by expensive investigations in the laboratory.

In the past two decades numerous management models for different standing waters have been developed on the basis of state determination. These models were mainly developed ad hoc for specific waters, with geographical spectrum including marine ecosystems (Steele, 1974; Kremer and Nixon, 1978), tropical waters (Vinogradov, Menshutkin and Shushkina, 1972) as well as arctic lakes (Howell, Krutchkoff and Parker, 1978). Most models are eutrophication models for typical Central European and North American lakes and reservoirs. The morphometric spectrum covers the North European shallow lakes (NYHOLM, 1978), large lakes such as Lake Balaton (SOMLYODY, 1981; KUTAS and HERODEK, 1982) and large and deep lakes such as Lake Baikal (MENSHUTKIN et al., 1980) as well as the Great Lakes of North America (PARK, 1974; DITORO et al., 1975; THOMANN et al., 1975; CANALE et al., 1976; SCAVIA, 1980).

In the first period of development, by using different approaches was the main target a high degree of reality of the models. To attain this goal, the following trends were pursued, with emphasis being laid on:

1. Model composition and structure.
   This trend resulted in very complex models. The model MS.CLEANER (PARK, 1974) considers up to 10 vertical segments of the water with 40 state quantities.
2. Ecological processes influence on model behaviour.
   This trend resulted in considering essential functional groups of species and decisive internal control mechanisms (STRASKRABA, 1979; BENNDORF and RECKNAGEL, 1982).
3. Chemical and biochemical processes influence on model behaviour.
   This trend resulted in taking into consideration internal nutrient load from the sediment (KAMP-NIELSEN, 1978) and internal conversion processes of organic and inorganic matter (LEONOV, 1979).
4. Hydrophysical processes influence on model behaviour.
   This trend resulted in taking into consideration advection and diffusion processes (IKEDA and ADACHI, 1979; EBENHÖH, 1980; OKUBO, 1980; WATANABE, HARLEMAN and VASILIEV, 1983).
5. Evolutionary processes influence on model behaviour.
   This trend resulted in taking into consideration processes of self-organization and adaption in aquatic ecosystems (IVACHNENKO, KROTOV and VISOTSKY, 1979; RADTKE and STRASKRABA, 1980).
6. Realism of exogenous variables and parameters of the model.
   This was achieved by discretization of historical input data for the water and year considered with a possibly high temporal resolution power, by the actual determination of water specific parameters and by preserving biological and biochemical parameters obtained by independent experimental investigations (BENNDORF, 1979b; RECKNAGEL, 1980).
7. Calibration of model parameters on a specific water.
   This trend was followed by the majority of model authors and generally, results in the „fixing" of the models descriptive validity for one water and a standard situation.

At the present stage of development, the generality of models must be increased in order to obtain predictive models. This requires a compromise between the trends mentioned above which can be characterized as follows:

— „Optimum" model composition and structure for the actual purpose of applications (DE WIT and RABBINGE, 1979) embodying only the essential conserved quantities of the system as state variables in the model and approximating self-organization of the system by conditional selection of alternative state variables in the model.
— „Maximum" consideration of internal control mechanisms of the system in the model (STRASKRABA, 1979) by paying most attention to a realistic description of biochemical

and ecological processes for regulation of the matter and energy flow between the state quantities.

- „Maximum" consideration of measured data for exogenous system quantities specifically for water, seasons and species, thus approximating hydrophysical processes.
- „Maximum" transparency, flexibility and compatibility of the model by means of appropriate data structures and computer program structures.
- Successive completion of the ecological model by appropriate hydrological, hydrophysical, terrestric models to obtain complex models which would also be appropriate for regional planning and control.

These aspects are supported by results obtained from direct comparisons of applications of different models to specified waters. In comparing the complex model MS.CLEANER (Park, 1974) and the very simple model AQUAMOD (Straskraba and Dvorakova, 1977) which has four state variables it was discovered that both models yield qualitatively similar simulation results for the state variables phytoplankton and orthophosphate of the Slapy Reservoir in Czechoslovakia (Straskraba, 1982).

In the paper of Walters, Park and Koonce (1980), the application of a „hypothetical" eutrophication model of a mean complexity to six lakes of the IBP was discussed. The spectrum of the lakes covers the hypertrophic Lake Wingra (USA) and Lake George (Uganda) and the oligotrophic Vorderer Finstertaler See (Austria).

The model parameters were obtained by independent experimental investigations and were maintained. Meteorological and hydrological input data were generated by empirical functions. Authors concentrate on the qualitative assessment of the seasonal dynamics of state quantities which are estimated as being realistic. However, quantitative comparison between the simulated and the measured state quantities revealed considerable deviations, which partly amounted to orders of magnitude.

An interesting clue in developing general eutrophication models is given in the paper of Swartzman (1979). On the basis of a so-called „simulation library" different specific models of North American lakes were used for comparing studies of the Lake Ontario. The comparison focused on evaluating reality of the rate variables of the models for those processes which decisively determine dynamics of the nutrients and the plankton components. In this way, the most credible process equations should be filtered from the different specific models in future to synthesize them into a „universal" model.

In developing the eutrophication model SALMO (Benndorf, 1979b; Recknagel, 1980) the first objective was to create a general and flexible management model for a wide class of standing water ecosystems. In the following sections, model validity and applicability will be discussed.

2.3. Computer concepts

For system theorists the computer is a tool, as basic and essential as the microscope is for biologists. Either of this tools enhances enormously the ability of the human being in a particular area.

G. J. Klir

The computer is the connecting link between the modeling process and the simulation process. In the simulation process, an appropriate computer program solves the model by numerically tracing the level of state or output variables through the control time or space under consideration. Analog computers as well as digital computers can be used for simulation. While the *analog computer* a priori permits the parallel processing of information, the *digital computer* is appropriately used in sequential processing.

In the *hybrid computer*, the hardware of both computer classes is integrated to enable a synthesis of parallel and sequential processing. Because of its almost unrestricted programmability, the digital computer is predestinated to simulate complex systems. However, the conventional sequential processing is often disadvantageous, because real-time simulation is unfeasible and hence because of increasing number of difficulties computing time proves as a limiting factor. An alternative can be found in the new architectures of the digital computer which are based on microprocessors tailored to an increased data throughput. Using the pipeline and parallelization principles, the so-called *parallel processor systems* acquire high computational speeds. By the pipeline principle instruction processing within a processor is speeded up by the timely overlapping of partial phases of instruction processing. Using the parallelization principle, simultaneous calculations on parallel processing units are aimed at. Simulation by parallel processor systems opens up new prospects in real-time simulation and control of large complex systems (cf. SCHMIDT, 1980; PIMENTEL, 1983; DEKKER, 1984). SUDO et al. (1982) already reported on results of digital computers consisting of 10^5—10^6 parallel processing units. In addition, the explosive distribution of *personal computers* promotes to apply digital computers for simulation in an interactive manner. These are at present 8, 16 or 32 bit computers but shortly it will be 64 bit computers for which cheap and large, directly addressable memory is available.

In the following sections simulation refers exclusively to the application of digital computers.

According to KUO and KAISER (1966) the coding of a model for simulation should adhere to the following principles: evolutionary, compatible and modular programming. *Evolutionary programming* means that the program author creates an algorithm which is self-assemblying an actual simulation procedure through selection of subroutines by means of call parameters. IVACHNENKO (1982) has brought to perfect this principle by the so-called self-organisation of models. Here, in the sense of a specific algorithm of computer artificial intelligence the computer itself can find the best unique model for control. *Compatible programming* means that the resulting program can coexist, communicate, and interact with other programs in the computational environment. For this purpose, a universal program language such as FORTRAN, BASIC or PASCAL should be used and program interfaces should be designed timely. The principle of *modular programming* is based on the coding of subroutines corresponding with components or subsystems of the modeled system. From the view of the simulation process functional groups of such subroutines can be collected to modules which make it possible to build a library of simulation modules (cf. SWARTZMAN, 1979). Modular programming allows the user to easily modify the program and it is necessary for evolutionary programming.

In adhering to these principles consequently, a high degree of transparency, flexibility and time efficiency of the computer program is guaranteed. Transparency is a very important factor in program verification for discovering coding artifacts. These principles are also supported by using appropriate simulation languages which were developed for continuous problems (e.g. CSMP) as well as for discrete problems (e.g. GPSS).

Generally, four simulation approaches can be distinguished. These are represented in Fig. 12.

The *scanning method* (Fig. 12a) requires an equidistant discretization of the control

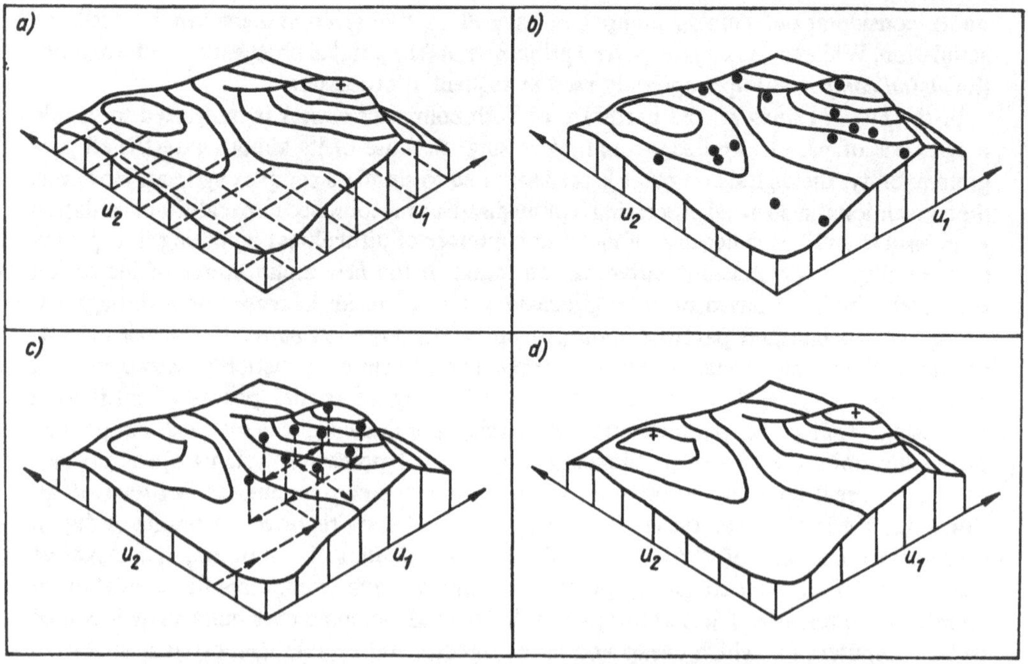

Fig. 12: Principal simulation approaches within the control space defined by the control variables u_1 and u_2.
a) Scanning method
b) Random method
c) Heuristic method
d) Method of steepest rise.
After LOUCKS, STEDINGER and HAITH (1981)

space. In Fig. 12a, two control variables u_1 and u_2 are taken into consideration, and the scan is registered by the $(u_1 \times u_2)$-matrix. By simulation according to the $(u_1 \times u_2)$-matrix, a relatively complete surface of system behaviour in the control space considered is provided and local optima as well as the global optimum can be registered. Accuracy and strength of the scanning method depend on the resolution power of the $(u_1 \times u_2)$-matrix, i.e. on the u_1 and u_2 selected. The system behaviour for arbitrarily selected values of the control variables is simulated by the *random method* (Fig. 12b). The probability to meet the local or global optimum depends on how often the simulation is repeated with new random values. The heuristic method (Fig. 12c) reduces corrections of the course within the control space to success or failure of previous simulations. This method results in a systematic restriction of the control space where an iterative approximation to local optima or to the global optimum is possible.

For the *method of steepest rise* (Fig. 12d) partial derivations of the objective function according to the changes of control variables are necessary. In this way, the rate of change, which maximizes the target function can be determined. As the partial derivatives can not generally be solved analytically within the limits defined for the control variables the region of the steepest rise is approximated by simulation.

It is possible to search for a global optimum by combining simulation approaches, e.g. by linking the scanning method and the method of steepest rise (see Chapter 5). To combine simulation methods can be advantageous, as it is pointed out later in Section 6.3. However, the computing time required will always be a limiting factor.

2.4. Control concepts

... a bee can put to shame a human master builder in the manner in which it forms its alveolus. But what the worst master builder marks out against the best bee is that he has constructed the alveolus in his head before he builds it in wax.

K. Marx

In general, when speaking about theory of control, we proceed from the fact that a system always consists of input variables which are made up of the control vector $\boldsymbol{U}$ and the disturbance vector $\boldsymbol{X}$ in which the noncontrollable input variables are included. By means of suitable control variables, negative effects of disturbances or noise on system behaviour are to be minimized. In accordance with the state description the system's control can be interpreted as follows. The system behaviour within a control period T is continuously represented by state vector $\boldsymbol{Z}$:

$$\dot{\boldsymbol{Z}} = \boldsymbol{f}(\boldsymbol{Z}, \boldsymbol{U}, \boldsymbol{X}, t),$$

where the initial condition $\boldsymbol{Z}(0) = z_0$ must be known.

The output vector $\boldsymbol{Y}$ follows on correspondingly:

$$\boldsymbol{Y} = \boldsymbol{g}(\boldsymbol{Z}, \boldsymbol{U}, \boldsymbol{X}, t).$$

According to Williams (1982), the goal function of an optimum control of system behaviour can be expressed as follows: it is based on an evaluated set of control sequences, the value of which has to be maximized:

$$\max J = \int_{t_0}^{t_f} I(\boldsymbol{Z}, \boldsymbol{U}, t)\, \mathrm{d}t. \tag{1}$$

The function I denotes the optimality index whereas t_0 represents the initial time and t_f represents the final time of the control period T. The characterized general control problem can be denoted as *unsteady state control* as opposed to steady state control. In Sections 6.3 and 7.3 the procedure and results of an unsteady state eutrophication control of reservoirs are illustrated. Provided that all variables are considered to be constant in time, the problem can be seen to find a satisfactory or optimum equilibrium of the system. In this case it is the question of *steady state control*, where

$$\boldsymbol{f}(\boldsymbol{Z}, \boldsymbol{U}, \boldsymbol{X}, t) = \boldsymbol{0},$$

$$\boldsymbol{g}(\boldsymbol{Z}, \boldsymbol{U}, \boldsymbol{X}, t) = \boldsymbol{0}.$$

In Chapter 4 the procedure and results of a steady state control of waste stabilization ponds are represented.

From the strategical view, different approaches for optimal control of a system can be applied. However, two fundamental approaches are distinguished: open-loop control and closed-loop control. In open-loop control the control vector has only a time dependence $\boldsymbol{U} = \boldsymbol{U}(t)$. However, in the case of closed-loop control the control vector $\boldsymbol{U}$ requires to be a function of state vector $\boldsymbol{Z}$ of the system under consideration: $\boldsymbol{U} = \boldsymbol{U}(\boldsymbol{Z}(t))$. In Fig. 13 the most important appearances of these two control approaches are illustrated.

Open-loop control (Fig. 13a) is the simplest control strategy. The control variables are empirically defined and they influence the system by a constant amount. Conclusions are

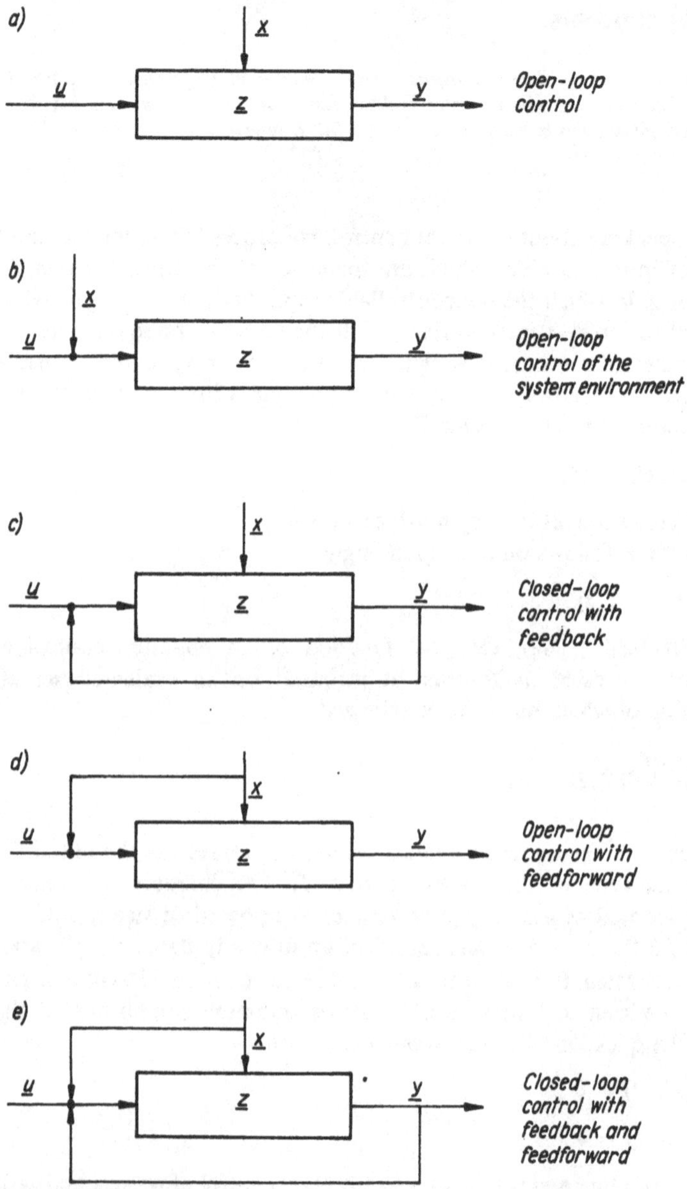

Fig. 13: Elementary control approaches

not drawn by means of the resulting output values to rearrange the control variables. The control variables are actualized in irregular distances on the basis of new experience. This control strategy is too sluggish to counteract random disturbances or suddenly occuring extremes in system behaviour.

Control of the system environment (Fig. 13b) is a special case of open-loop control and is characterized by the fact that the disturbances are already opposed in the system environment by appropriate control variables. Thus, in a simple way, an external preventative control is achieved. However, this type of open-loop control requires comprehensive knowledge of external disturbances and requires the application of control possibilities for its opposition before they enter the system. *Closed-loop control with feedback* (Fig. 13c) is based on continuous comparison of system output variables with the control target. The deviations which occur are the basis for corrections of the control variables. Effects of disturbances on system behaviour are subsequently compensated when this control strategy is applied correctly.

Open-loop control with feedforward (Fig. 13d) is based on the human ability to identify disturbances and to assess their consequences for system behaviour, before they affect the system. Thus, an internal preventative control is obtained in which the system is more resistant in the face of external disturbances. This control strategy requires identification and continuous measuring of external disturbances as well as knowledge of appropriate control variables. It is a disadvantage that comparison of output variables with the control target is impossible.

Closed-loop control with feedback and feedforward (Fig. 13e) combines to a certain extent the advantages and at the same time compensates the disadvantages of the two fundamental control approaches. The feedforward elements can prevent to system disturbances which are identified in time, whereas the feedback elements can reduce the effects of imperfections of feedforward elements and of disturbances which have not yet been registered.

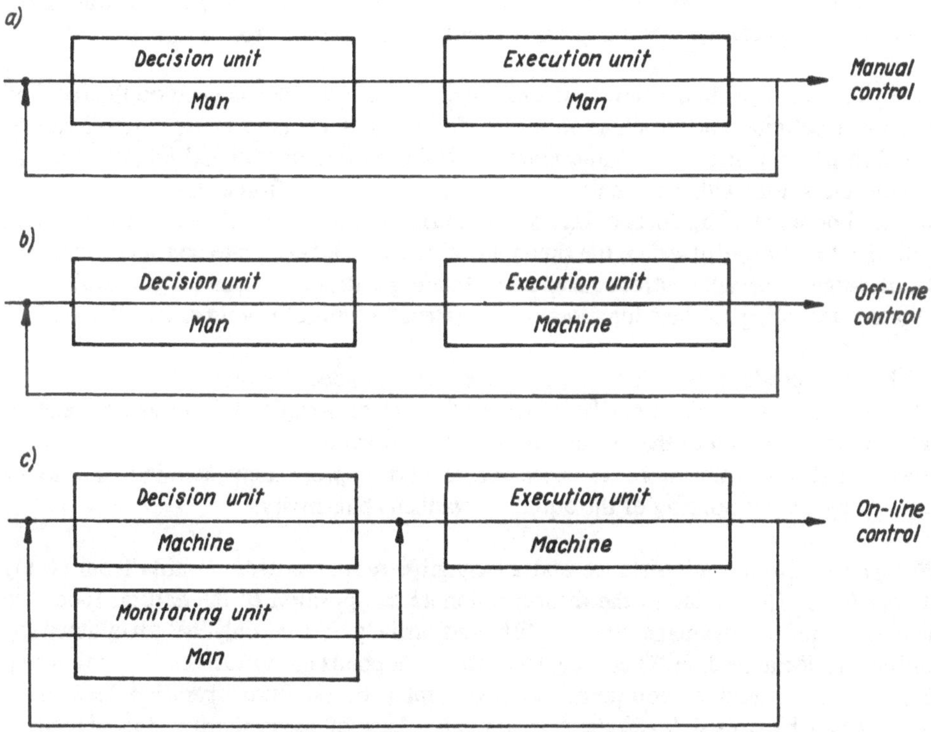

Fig. 14: General control equipments

Adaptive feedback-feedforward control represents a relatively universal and idealized concept in which the control concept is able, by subsequent restructuring, to carry out continuous modifications of the goal functions, of control variables and of perceptions of disturbances. So far, this control has not been realized in practice, however, it competes with evolutionary systems in its ability for self-organization and adaptation.

Generally, a control equipment is subdivided into three elements: decision unit, execution unit and monitoring unit. According to the extent of incorporating computer and measuring techniques into control, manual, off-line and on-line control equipment are thus distinguished (see Fig. 14). In *manual control* (Fig. 14a) the decision unit and the execution unit are carried out by men. In *off-line control* (Fig. 14b) a machine is used in the execution unit whereas in *on-line control* (Fig. 14c) beside the execution unit the decision unit is also performed by a machine, namely the computer. The monitoring unit is always the responsibility of the person in charge and is only taken explicitely into consideration in on-line control. In the manual and off-line control the function of the monitoring unit is implied into the decision unit.

According to LANGE (1970), the efficiency of a definite control concept can be estimated under the following criteria:

1. The speed with which the control neutralizes effects of disturbances on the system
2. The precision with which the desired system behaviour is obtained by control
3. The reliability of a control system in preventing the system behaviour from collapsing
4. The stability of control to ensure a constant reduction of disturbances
5. The sensitivity of control to insignificant deviations in control variables or disturbances
6. The cost efficiency, i.e. obtaining an efficient control system from a limited budget.

Finally, for control, different time horizons are distinguished which depend on the required speed for a definite control situation and a definite system. *Operational control* can be realized in an on-line or an off-line manner. While on-line operational control requests the real-time status with time units of minutes or seconds, off-line operational control tolerates time units of up to one day. *Management control* refers to time units of weaks or months. In *strategic planning*, the time unit of one year is taken into consideration.

To investigate relevant control concepts for standing waters it is ingeniously to distinguish the waters according to their functions with respect to natural resources as follows:

1. Fish ponds produce fish which use resource water of adequate quality
2. Waste stabilization ponds purify resource water from organic and inorganic loads by maximum utilization of the bioactivity of aquatic organisms
3. Lakes and reservoirs store resource water and appropriately stabilize the water quality by proper utilizing of the aquatic organisms bioactivity.

Fish ponds have a low resistance and a negligible resilience against shifts from steady state positions. This is due to the disproportionate composition of the natural food web which is of more advantage for the fish and usually it can only be maintained by additional artificial feeding. Therefore, to control fish ponds an operational control, which can adequatly prevent or counteract averages, must be put into operation seasonally. Fish kills can be caused (a) if there is a sudden lack of oxygen, (b) if the pH-values are too high and/or (c) if there are external or internal toxic pollutants or pathogens

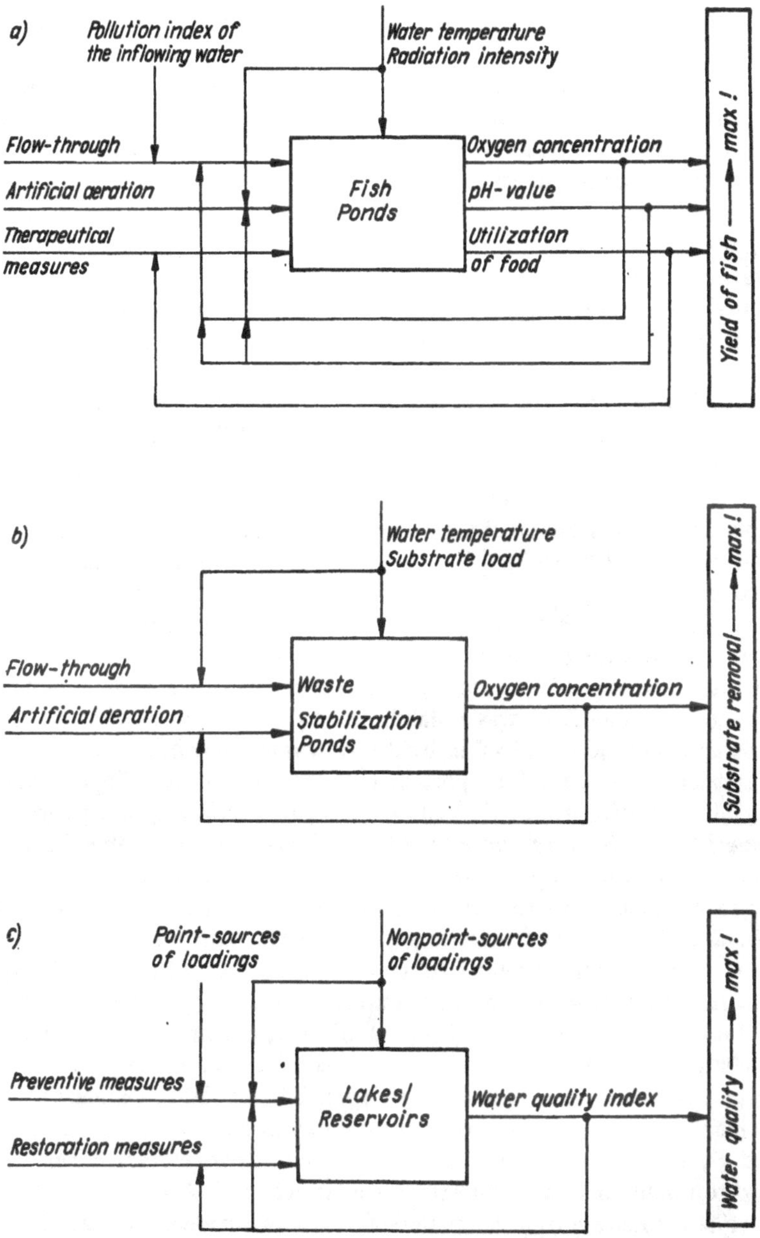

Fig. 15: Adequate control concepts for standing waters.
a) Fish ponds
b) Waste stabilization ponds
c) Lakes and reservoirs

present in the water. In all of these cases, a considerable burden is placed on the state's finances. In the ideal control concept for fish ponds, the feedforward and beedback elements of the environmental control should be combined. In environmental control, a continuous water inflow, which occurs only when adequate water pollution indices do not exceed definite limit values, which may give rise to toxic dangers for the fish, must be guaranteed. The artificial aeration should be controlled by means of feedforward elements which depend on the water temperature and on the period of radiation. Depending on the measured O_2 concentration or on the pH-value, the feedback elements must operatively release an artificial aeration and/or an increase in the flow-through. In observing toxic pollutants or pathogens of fish, i.e. a decrease in feed intake, appropriate preventative measures have to be implemented by feedback elements (i.e. treatment by biocides).

Waste stabilization ponds are extensively based upon the lowest joints of a biological food web such as bacteria, ciliates, phytoplankton which because of their high rates in growth and selfreproduction have a distinct self-regulation and adaptability which increase the system's resistance. As a result, by using an adequate control concept, only for waste stabilization ponds maximization of the self-purification performance within a management control can be pursued (see Fig. 15b). Water residence time is optimally controlled by feedforward elements dependent on external organic or inorganic load and on water temperature. Thus, the substrate removal performance is maximized. In addition, the substrate removal performance, especially of waste stabilization ponds is seasonally increased by artificial aeration depending on oxygen measurements in water. In Chapters 3 and 4 a substrate removal model of waste stabilization ponds is illustrated and is very well suited to be used as a decision tool in feedforward management control.

For *lakes and reservoirs*, a well-proportioned biological food web of optimum bioactivity concerning effective performance of aquatic organisms — according to the uses of waters — is a prerequisite for a high resistance to maintain a stable water quality. However, when the resistance is exhausted by permanent high external loads, the waters are not sufficiently resilient to revert back to the former stable steady state position. Often, the shifts of the system into a new steady state position are irreversible. Therefore, in case of lakes and reservoirs, the combination of control of the environment with feedforward and feedback elements in the frame of a management control or strategic planning is to be aimed at (see Fig. 15c). In the sense of an external preventive control with respect to point sources pollution, control of the environment should be given priority. Such point source pollutions are discrete, localizable discharges to which industrial and municipal waste water treatment plants, pre-dams, waste water from agricultural plants and dumping grounds be assigned. By means of appropriate feedforward elements an internal preventative control with regard to pollution from nonpoint sources is achieved. According to KRENKEL and NOVOTNY (1980), pollution from nonpoint sources is subdivided into natural and anthropogeneous sources. Weathering processes of minerals, soil erosion of areas without closed vegetation cover, inflow of residual matters of natural vegetation and atmospheric inflow belong to natural sources. The most important anthropogeneous nonpoint sources are mineral fertilizers, chemical protectants and pesticides in agriculture and forestry.

In comparing system output quantities and specific limit values of water quality indices adequate to uses, an assessment of control efficiency is carried out by the corresponding feedback elements. As a result of this assessment, if possible, corrections of control variables can be deduced. When extreme internal loads occur, by feedback elements a restoration strategy to control internal loads will be induced. Among internal loads there are high release rates of mineral nutrients from the sediment or a high oxygen consumption in the sediment due to a considerable organic load of sediment.

The use of the applied systems analysis to assist the management control and strategic planning of lakes and reservoirs is subject of Chapters 5, 6 and 7. In these Chapters the above mentioned control concept is put into practice.

3. An Input/Output Model for Waste Stabilization Ponds

Waste stabilization ponds are artificial waters used as reactors for biological waste-water treatment. In contrast to natural waters, an exceptional high bioactivity is aimed at (UHLMANN, 1979a). In warm climates, waste stabilization ponds represent the most important method for biological treatment of domestic effluents to the extent that the water can be safely re-used for irrigation and pisciculture. In Europe and in North America, ponds in many cases serve to treat effluents from small communities or are used for an advanced treatment stage.

The performance of waste stabilization ponds is determined by the biochemical removal of organic matters. The optimum number n of basins „connected in series" and the total residence time t of water in the pond system are used as control variables to maximize the removal performance. Supposing that each single basin of the pond system is considered to be a completely mixed reactor, similar to chemical and microbiological engineering the general equation of mass balance for mixing reactors is taken as a basis:

$$\frac{dS}{dt} = DS_0 - DS - k_1 S, \tag{2}$$

where S_0: concentration of the substrate inflow in mg BOD_5/l
S : concentration of the substrate outflow in mg BOD_5/l
D : dilution rate of water in 1/d
k_1 : biochemical removal rate of substrate in 1/d.

Supposition of the steady condition with $\frac{dS}{dt} = 0$ restricts equation (2) to the consideration of the mean seasonal behaviour of a waste stabilization pond and provides the reference point for the subsequent modeling. Thus, the results of equation (2) are as follows:

$$\frac{S}{S_0} = \frac{1}{1 + k_1 \bar{t}}, \tag{3}$$

where $\bar{t}$ corresponds to $1/D$.

The total pond system can now be described as a completely mixed, continuously stirred tank reactor (CMTR) in series as proposed by MARAIS (1966):

$$\frac{S}{S_0} = \frac{1}{\left(1 + \frac{k_1 \bar{t}}{n}\right)^n}. \tag{4}$$

Applying equation (4) to the optimum control of the substrate removal performance S/S_0 requires to calculate the biochemical removal rate k_1 in relation to the substrate load L, the mean water temperature T and the total residence time $\bar{t}$ of the water:

$$k_1 = f(L, \bar{t}, T), \qquad (L, \bar{t}, T) \in \boldsymbol{L} \times \bar{\boldsymbol{t}} \times \boldsymbol{T}$$

The determination of the transfer function f was made by input/output description of the system.

Already in Section 2.2.1, we referred to the key position of a reliable and representative data base for a successful input/output description. In case of the present problem, the data base was generated by means of a laboratory simulator which allowed an experimental simulation of a relevant matrix of conditions. From investigations of similarity which referred to biological, biochemical and hydrophysical criteria the applicability of the laboratory simulator for this purpose was thus confirmed. Definite information on validity, construction and operation of the laboratory simulator is included in UHLMANN (1979b).

In order to experimentally simulate the biochemical removal rate k_1 the matrix of conditions was defined as follows:

$$\boldsymbol{L} = \{1, 5, 25, 125\} \qquad \text{mg BOD}_5/\text{ld}$$

$$\bar{\boldsymbol{t}} = \{5, 10, 20, 40\} \qquad \text{d}$$

$$\boldsymbol{T} = \{4, 10, 20, 30, 40\} \qquad {}^\circ\text{C}$$

$$\boldsymbol{L} \times \bar{\boldsymbol{t}} \times \boldsymbol{T} \rightarrow \boldsymbol{k}_1, \quad \text{card} \quad (\boldsymbol{L} \times \bar{\boldsymbol{t}} \times \boldsymbol{T}) = 80 .$$

In the following, the resulting data file was taken as the basis of an iterative multiple regression analysis, in which equation (5), which can be used to calculate k_1, was determined:

$$k_1 = \bar{t} \uparrow [-1/(1.39 + 1.3/T + (0.06 + 0.05 \cdot T)/L)] \\ /[0.32 + 10.27/T + 1/(0.25 + 0.47/T)/L] . \qquad (5)$$

In order to check the validity of the complete model (substitution of k_1 in (4) by (5)) the substrate removal performance S/S_0 for different real-existing ponds was calculated and compared with the measured values. A sensitivity analysis was also carried out (see also UHLMANN and RECKNAGEL, 1982; UHLMANN et al., 1983). In Fig. 16a, the comparison for calculated and measured data is represented for five large scale waste stabilization ponds in the German Democratic Republic, where geometric means of the BOD in the effluent were used for different seasons (spring, summer, autumn, winter) and for different sampling points (outflow of basin 1, 2 and 3, respectively). Fig. 16b illustrates the comparison for waste lagoons in tropical climates (Cuba, Brazil and India) where measuring data were taken from literature (BRADLEY and SILVA, 1977; MARA, 1979; RAO, 1972). The enveloping curves of Figures 16a and b represent the 95% confidence limit. In both figures the values measured well agree with those calculated in spite of considerable differences of the systems considered with respect to the size of the basins and the climatic conditions. The relatively high rate of correspondence for tropical ponds is explained by the high temperature ranges up to 40 °C considered in the data base.

The sensitivity analysis in Fig. 16d makes evident the „compatibility“ of the model with different sizes of basins, where a low sensitivity of the model against changes

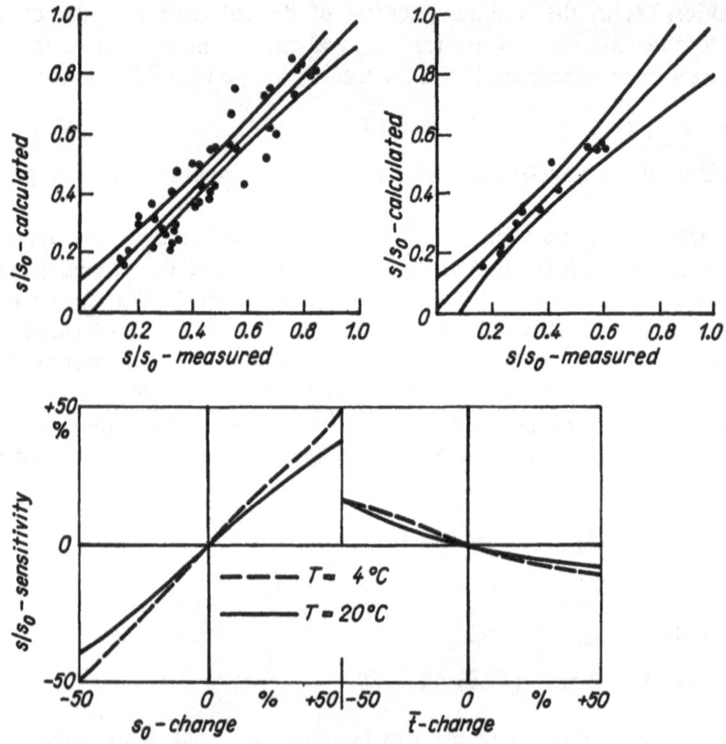

Fig. 16: Validation of the substrate removal model for waste stabilization ponds. After UHLMANN and RECKNAGEL (1982).

a) Comparison of calculated and measured data for the substrate removal performance S/S_0 of ponds in the GDR

b) Comparison of calculated and measured data for the substrate removal performance S/S_0 of ponds in Cuba, Brazil and India

c) Model sensitivity against changes of the substrate inflow concentration S_0 by $\pm 50\%$

d) Model sensitivity against changes of the total residence time $\bar{t}$ by $\pm 50\%$

of the total residence time $\bar{t}$ up to $\pm 50\%$ can be observed. As expected, the sensitivity of the model against changes of the substrate inflow concentration S_0 up to $\pm 50\%$ is increased (see Fig. 16c).

Along with the validation, the substrate removal model has proved to be predictive for waste stabilization ponds in the range of substrate inflow concentrations from 200 to 800 mg BOD_5/l, of water temperatures from 4 to 40 °C and of mean total residence times up to 120 d.

4. Model-based Control of Waste Stabilization Ponds

As outlined already in Section 2.4 a management control is necessary for waste stabilization ponds which is aimed at taking advantage of their biological self-purification capacity in an optimal manner. To meet this target, the substrate removal performance S/S_0 has to be optimized, where the control variables mean residence time $\bar{t}$ of the water and number n of basins in series are considered. The substrate removal model proposed in Chapter 3 for solving the optimization problem has proved to be suitable.

In Fig. 17 the off-line approach of the model-based control of optional waste stabilization ponds is represented according to RECKNAGEL and UHLMANN (1986). Provided that the measured input vector X and the measured output Y are available, the optimum control trajectories of vector U can be determined, thus bringing about a removal performance for the approximation of the limit values of the substrate outflow concentration according to the national Standard of Waste Water Treatment. The corresponding goal function of the optimization problem reads:

$$\text{abs}\,(S_L - S(S_0, T, \bar{t}, n)) \rightarrow \min_{(\bar{t},\, n)}$$

where $0 < S_0/\bar{t} \leqq 5$, $n = 1(1)\,5$
and S_L: seasonal limit value of the substrate outflow concentration mg BOD_5/l as defined in standards for waste water treatment.

According to Fig. 17 the approach is based on computer simulation, where three steps are distinguished: model initialization, simulation of control diagrams and optimization of the control trajectories.

To make transparent the steady state behaviour of ponds within an adequate control space, seasonal control diagrams are simulated. For simulating control diagrams the scanning method is combined with the method of the steepest rise according to Section 2.3.

The control diagrams in Figs. 18 to 20 were generated for a fictive pond system, they were used with two or three basins and were referred to the seasonal limit values S_L of the substrate outflow concentration as defined in the GDR-standard (TGL 28722). The control space considered was defined by mean water temperatures in the range $T = 4(2)\,40$ °C (X-axis) and substrate inflow concentrations in the range $S_0 = 200(20)$ 400 mg BOD_5/l (Z-axis). The left-hand column of figures was referred to two basins and the right-hand column of figures was referred to three basins under consideration. In each case the percentage of substrate removal performance (figures above) and substrate outflow concentration (middle of the figures) corresponding to the optimum values of mean residence time of the water (figures below) were represented. The references of

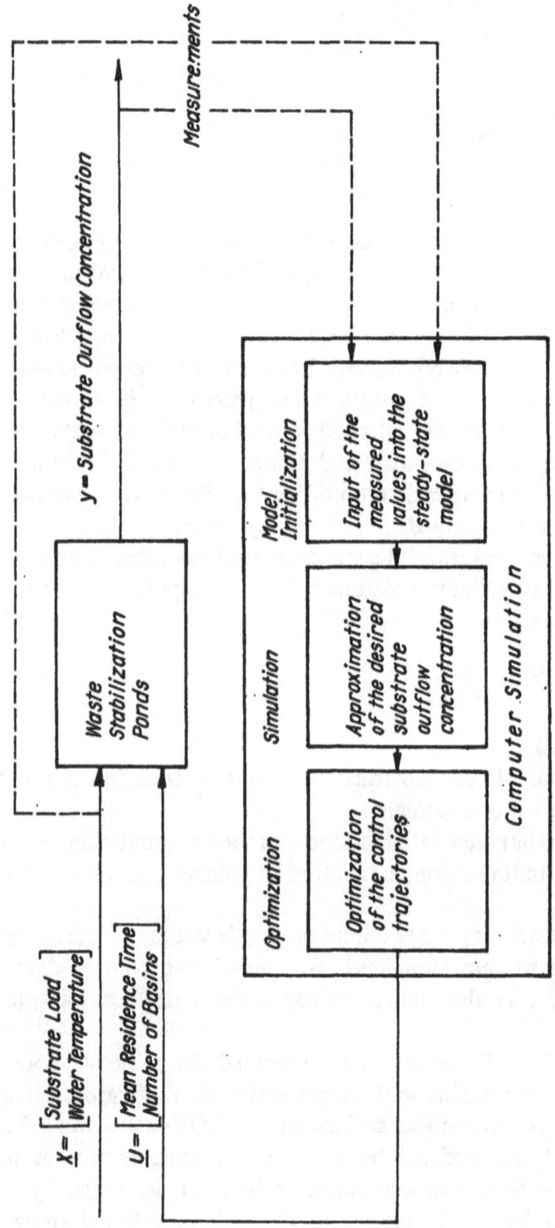

Fig. 17: Off-line approach of the model-based control of waste stabilization ponds

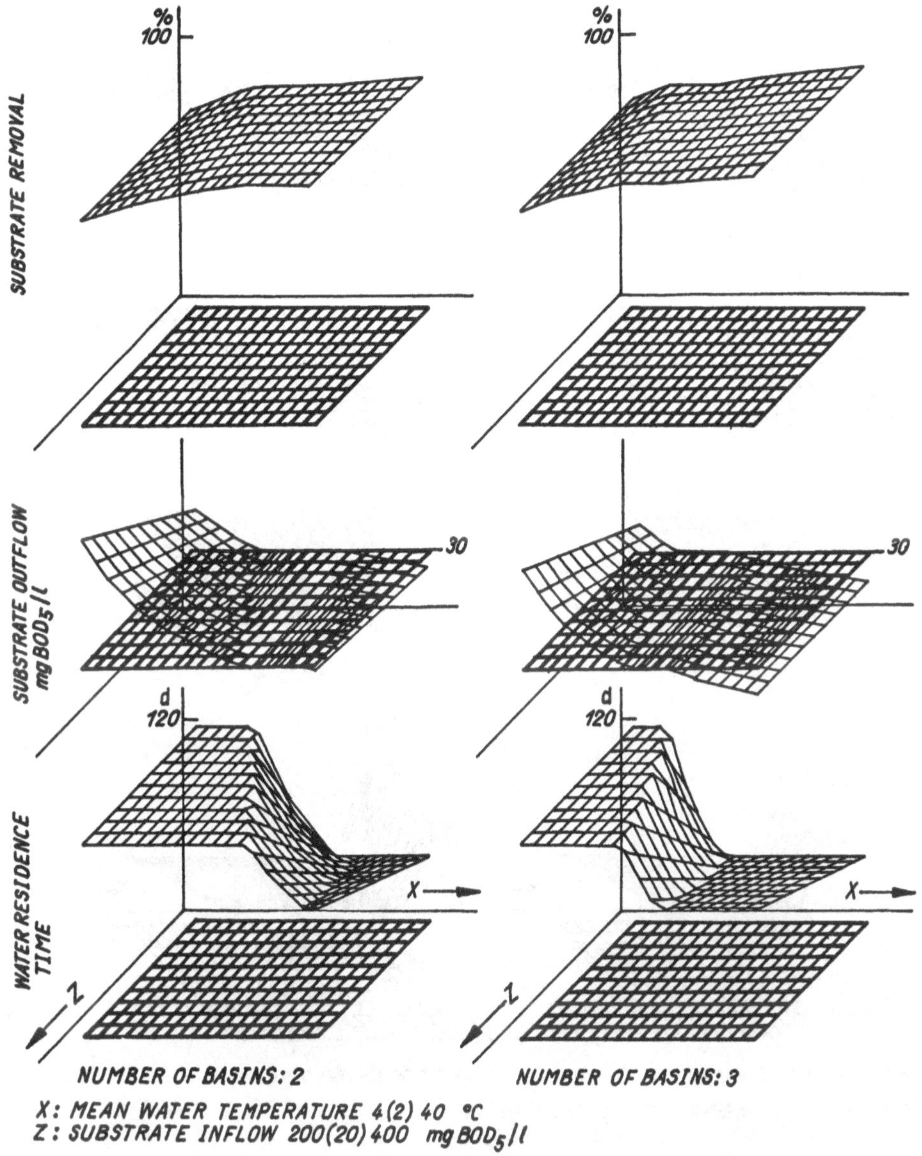

Fig. 18: Control diagram for a fictive waste stabilization pond referred to the limit value S_L = 30 mg BOD_5/l of the substrate outflow concentration in summer

the seasonal limit values S_L were plotted into the figures of substrate outflow concentrations.

It is obvious from Fig. 18 that in summer the limit value $S_L = 30$ mg BOD_5/l with two basins (Fig. 18 left-hand) is steadily maintained or lower by optimum control of the residence time at water temperatures above $T = 25$ °C. At temperatures below $T = 25$ °C the limit value S_L is increasingly exceeded in spite of using the maximum residence time of $\bar{t} = 120$ d with increasing substrate inflow concentration. By connecting

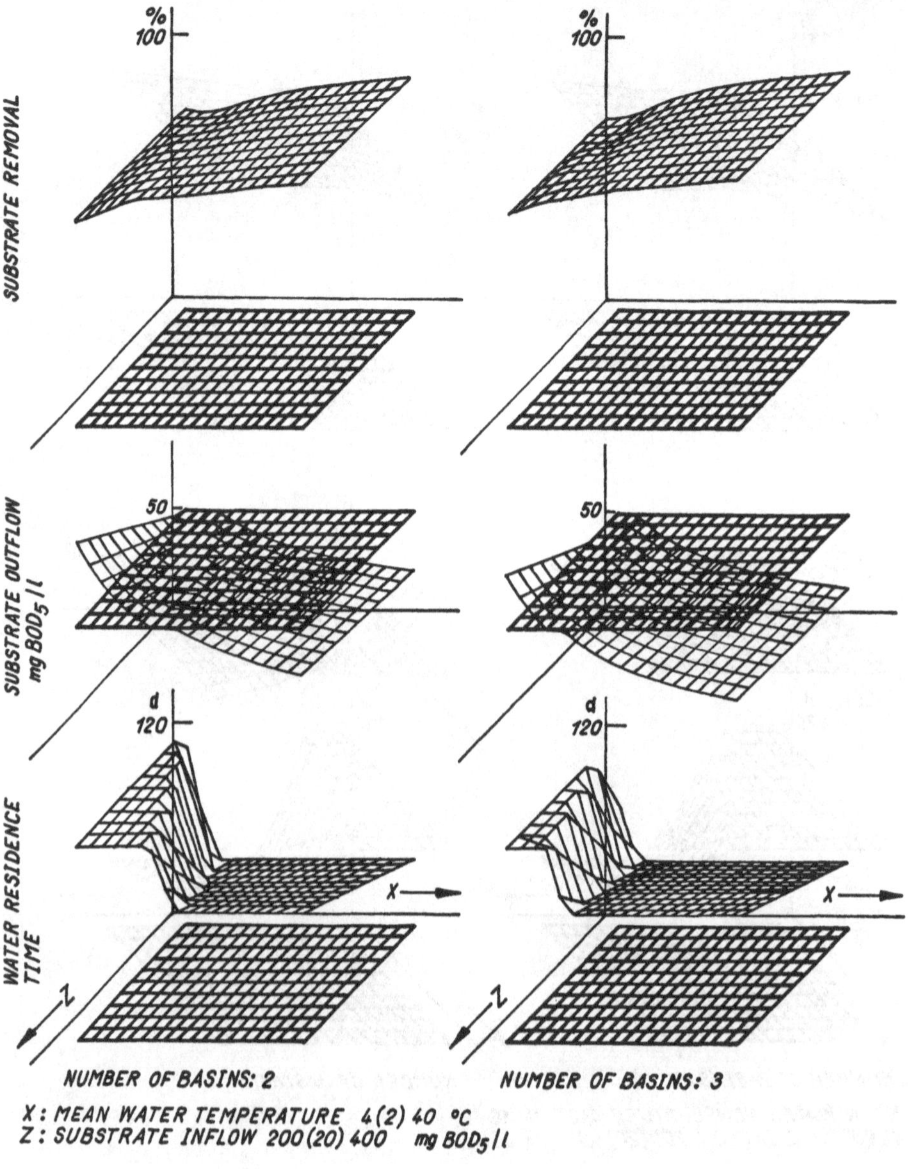

Fig. 19: Control diagrams for a fictive waste stabilization pond referred to the limit value $S_L = 50$ mg BOD$_5$/l of the substrate outflow concentration in spring or autumn, respectively

with a third basin (Fig. 18 right-hand), the temperature limit for the steady maintenance of the limit value S_L is reduced to $T = 20$ °C; below that, S_L is exceeded with increasing S_0 in spite of the maximum residence time $\bar{t} = 120$ d. In both cases, substrate removal performance is in the range of 70 to 96% (Fig. 18 above).

In Fig. 19, the spring or autumn aspect, with the limit value $S_L = 50$ mg BOD$_5$/l is considered respectively. The temperature limit to steadily maintain the value S_L or to exceed it for two basins is at $T = 16$ °C (Fig. 19 middle of the left-hand) and for

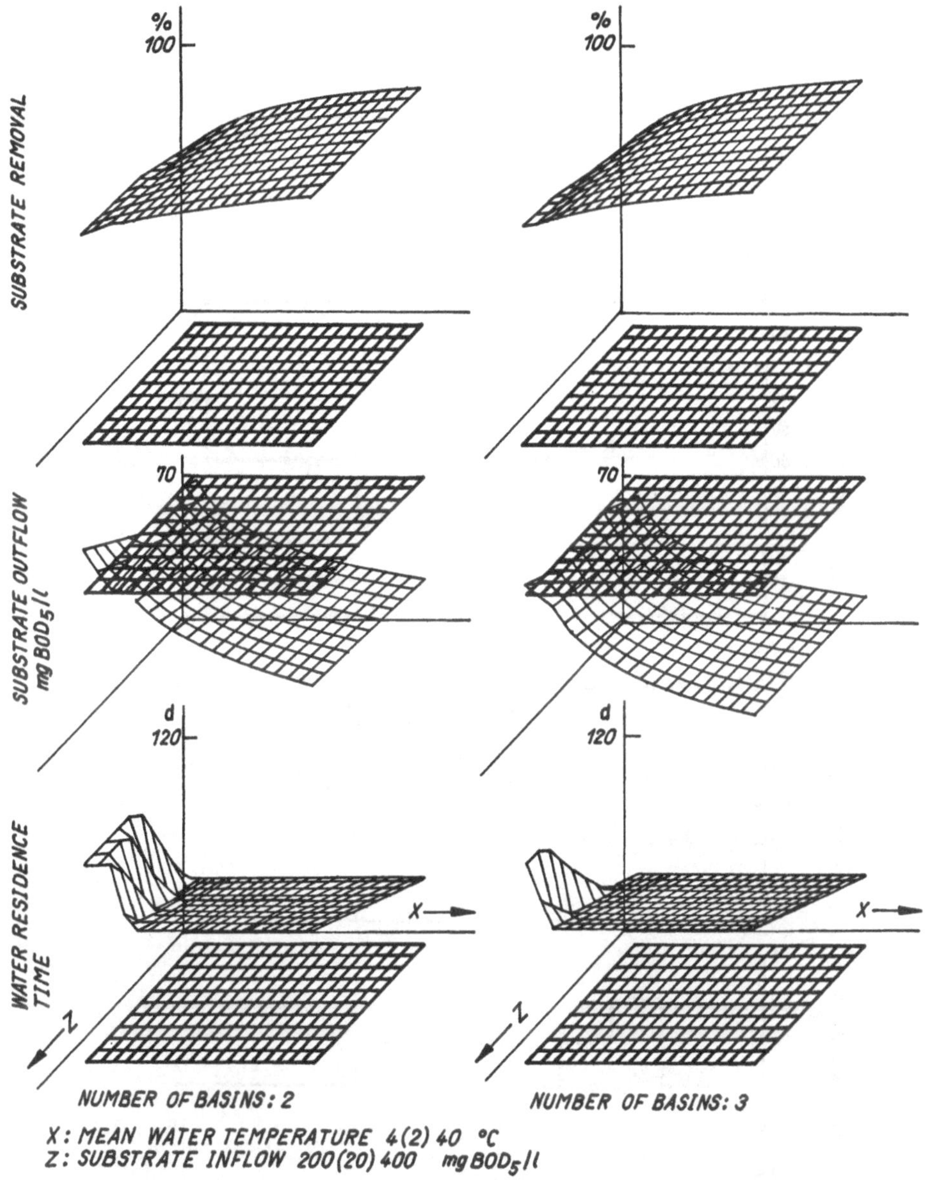

Fig. 20: Control diagrams for a fictive waste stabilization pond referred to the limit value S_L = 70 mg BOD_5/l of the substrate outflow concentration in winter

three basins it is T = 10 °C (Fig. 19 middle of the right-hand). Above these temperatures the system can be controlled by minimal residence times of $\bar{t} = S_0/5$ d (see Fig. 19 below). Here, substrate removal is in the range of 72 to 98% (Fig. 19 above), where higher temperature ranges are scarcely relevant.

In Fig. 20, the winter aspect with the limit value S_L = 70 mg BOD_5/l is represented. Here, in order to steadily maintain the limit value for two basins up to a substrate

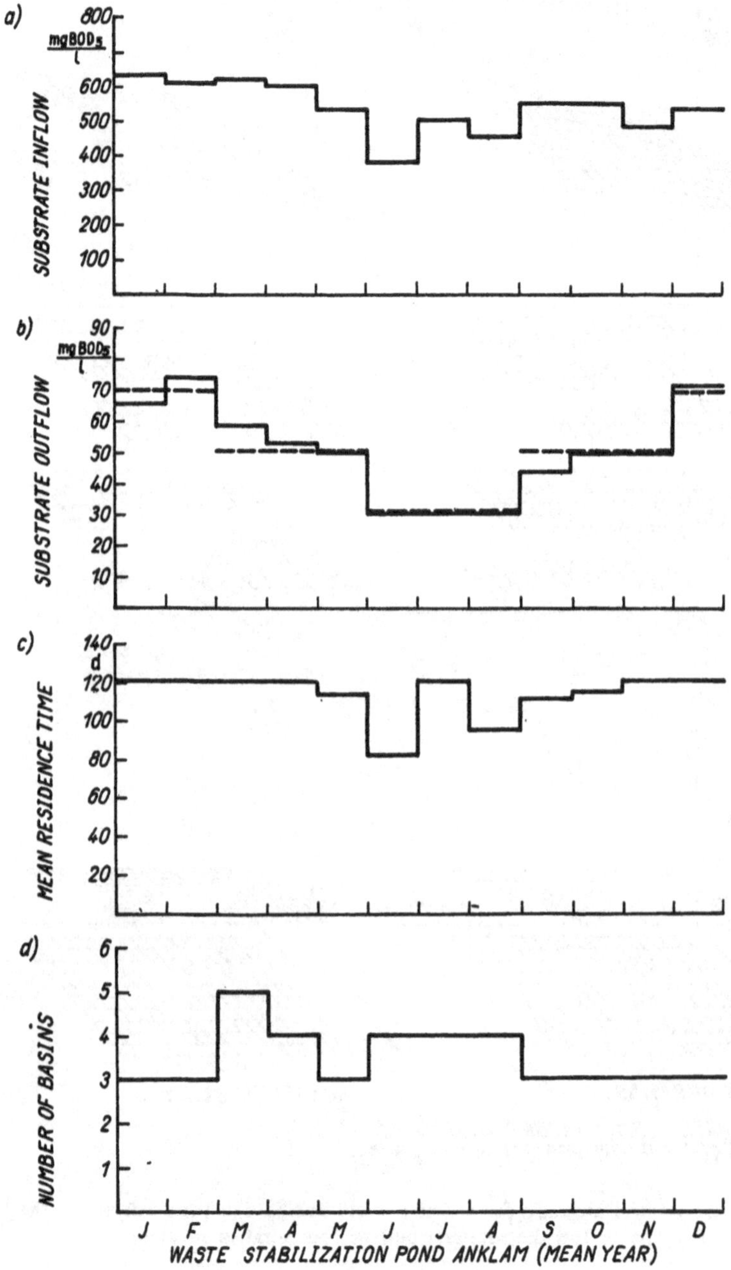

Fig. 21: Optimal control of the waste stabilization pond of a town with a population of about 25.000 people (mean year).
a) Monthly substrate inflow concentrations in mg BOD_5/l
b) Optimum trajectory of the substrate outflow in mg BOD_5/l (——— seasonal limit values)
c) Control trajectory of the mean residence time in d
d) Control trajectory of the number of basins.

inflow concentration of $S_0 = 300$ mg BOD_5/l the temperature is $T = 4$ °C and for more increased substrate concentrations is $T = 8$ °C (Fig. 20 middle of the left-hand). For a cascade of three basins, all connected with one another, a steady maintenance of the limit value S_L up to a substrate concentration of $S_0 = 360$ mg BOD_5/l at temperatures of $T = 4$ °C is secured by optimal control of the residence time. Increased substrate concentrations can only be compensated at temperatures above $T = 6$ °C by an optimum residence time (see Fig. 20 middle of the right-hand). Substrate removal is in the range of relevant temperatures of $T = 8$ °C between 60 and 75% (Fig. 20 above). Within the framework of a case study the approach was applied to the high-loaded waste stabilization pond of a town with a population of about 25.000 people. We proceed from the assumption that in future the pond system can be used for five basins and the mean residence time can be efficiently controlled. By using monthly mean values of input variables from the years 1972 to 1976 monthly control periods were assumed. In Fig. 21, a mean year for a pond system is illustrated. The substrate inflow concentrations (Fig. 21a), the resulting optimum trajectory of the substrate outflow (Fig. 21b) and the corresponding control trajectories of the mean residence time (Fig. 21c) as well as of the number of basins (Fig. 21d). As to be seen from Fig. 21b, the optimum seasonal limit values (dotted line) were maintained approximately as defined in the GDR-Standard for Waste Water Treatment (TGL 28722/01), notwithstanding that extreme inflow concentrations of about 600 mg BOD_5/l temporarily occur (see Fig. 21a).

Theoretically, the input/output model be appropriate for on-line control provided that input data were available from a continuously or periodically operating measuring system and provided that the controlling technologies were designed with sufficient sensitivity. In this case, the computer simulation steps in Fig. 17 would be performed on the spot by a microprocessor.

5. State-determined Model SALMO for Lakes and Reservoirs

The basic version of the model SALMO (Simulation by means of an Analytical Lake MOdel) was elaborated by BENNDORF (1979a) and RECKNAGEL (1980) and meanwhile further improvements have been made. According to Section 2.2.2 from the beginning a predictive model was sought for application as a flexible decision making tool in eutrophication control. For this purpose the essential components and the significant biological and chemical processes of aquatic ecosystems were incorporated successively. This bottom-up approach has proved to be suitable and will continue to be used.

5.1. Description of model

The model SALMO was designed as a two-layer model which would make the separate simulation of system quantities in the epi- and hypolimnion of stratified waters possible during the summer period. It is assumed that the concentrations of matter or biomass are homogeneously distributed within the water body under consideration and all calculations are referred to a unit volume of one cubic meter of water over a specified period of time.

In Fig. 22 for the present version of SALMO driving variables, state variables and their principal interactions are roughly outlined. As can be seen from Fig. 22, two nutrient components which can be limiting for primary production in waters can be simulated: dissolved orthophosphate and dissolved inorganic nitrogen. At present the phytoplankton component is subdivided into two state variables for different functional groups. In group 1 diatoms of the type Asterionella or cyanophycees of the type Oscillatoria can be considered selectively whereas in group 2 chlorococcales and diatoms of the types Scenedesmus and Cyclotella are contained permanently. For the state variable of the herbivorous zooplankton it was defined to consist of chrustaceans of the types Daphnia, Eudiaptomus and Cyclops. The system components fish and predatory zooplankton are included in an indirect manner by considering their influence on zooplankton mortality. In addition the component detritus can be simulated as an alternative food source for zooplankton but also dissolved oxygen in the hypolimnion during the summer period. Each of the described state variables is represented by an ordinary differential equation which is solved by the four-step Runge-Kutta method according to JENTSCH (1969).

The data requirements of the model are characterized as follows. As driving variables are considered: incident solar radiation, water temperature and temperature stratification, external loads of nutrients, water inflow, water outflow, mean and maximum water

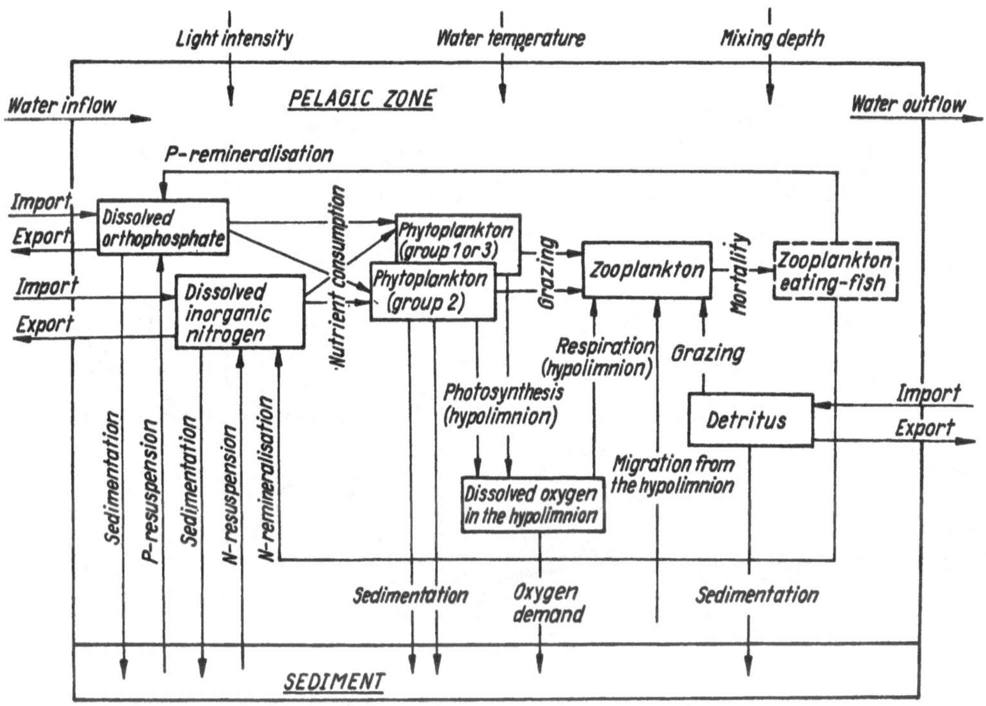

Fig. 22: Diagrammatic representation of the model SALMO with respect to the essential system quantities and their interactions

depth and volume of the water body. The data of the driving variables must be available for the specific water and year under consideration. They have to be prepared as annual timeseries by a decade-by-decade discretization using of mean values of historical measurement data. In this way it is possible to observe the model's response to real crest events in hydrological, meteorological and physical conditions of the year and water under consideration. In addition the water-specific parameter value of the maximum transmission of light must be determined to calculate the daily extinction coefficients of the water. The values of the other 117 parameters were obtained by independent field and laboratory experiments (Benndorf, 1979b) and are kept equally for different waters without calibrating.

To simulate a real water the initial conditions of the state variables also should be based on historical measurements and in accordance to the initial time of simulation. In Fig. 23 simulated trajectories of state variables of the running years 1975 to 1978 of the Saidenbach Reservoir are represented by using a historical data file for driving variables and initial conditions. As can be seen from the Fig. 23, annual-specific transient behaviours of the water are generated. In attempting to simulate the mean behaviour of a water by repeating the data file for driving variables by running for several years the model trajectories will pass through the transient behaviour and reach a steady-state oscillation (see also Park, 1974). This can be observed in Fig. 24 where for the simulation the data file of 1975 for the Saidenbach Reservoir was used repeatedly.

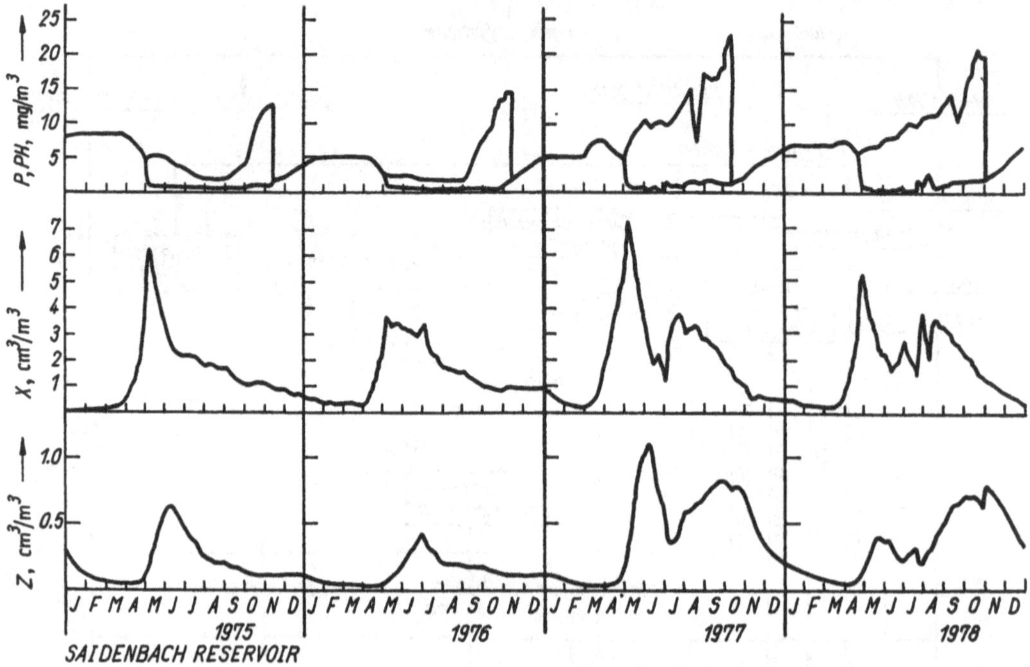

Fig. 23: Annual trajectories of the state variables dissolved orthophosphate (P, PH), total biomass of phytoplankton (X) and zooplankton (Z) of the running years 1975 to 1978 of the Saidenbach Reservoir by using historical input data files

Along with the modeling of SALMO special attention was paid to the construction of the process rates. According to BENNDORF and RECKNAGEL (1982) the process rates are based on the following ecological control mechanisms:

- Multiple resource kinetics determining phytoplankton photosynthesis and zooplankton grazing as proposed by STRASKRABA (1979)
- Different dependencies on the temperature of phytoplankton photosynthesis and respiration
- Sedimentation of particulate nutrients in faecal pellets of zooplankton
- Nutrient remineralization by zooplankton and fish
- Zooplankton feeding on phytoplankton
- Fish feeding on zooplankton
- Intrapopulation competition for nutrients and food to determine phytoplankton photosynthesis and zooplankton assimilation
- Decrease of food utilization with increasing food intake by zooplankton
- Self-shading by phytoplankton
- Food preferences of zooplankton related to various food species and detritus
- Egg development of zooplankton depending on temperature.

In Fig. 25 the Forrester notation was used to make evident the processes considered in SALMO for determining state variables. The legend of symbols which are used in Fig. 25 is given in Tab. 1. A more detailed structure diagram inclusive list of symbols is represented in BENNDORF, KOSCHEL and RECKNAGEL (1985). Corresponding to the Forrester diagram in Fig. 26, pages 1 to 7, the formal transfer functions of state and rate variables

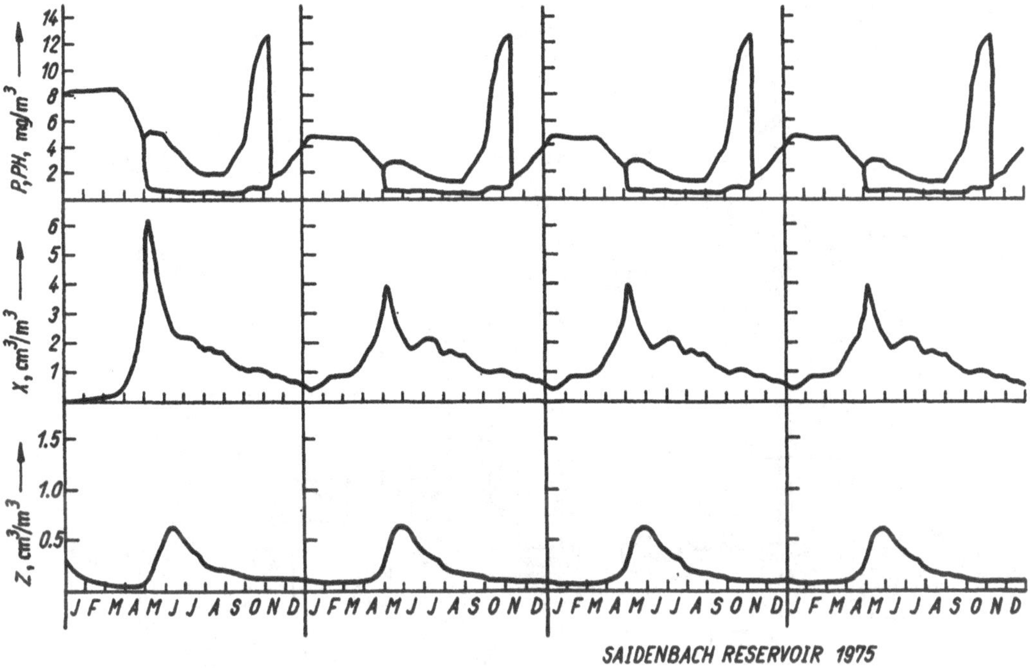

Fig. 24: Mean trajectories of the state variables dissolved orthophosphate (P, PH), total biomass of phytoplankton (X) and zooplankton (Z) by repeating of the input data file of the year 1975 of the Saidenbach Reservoir

as well as the attached annual trajectories are documented by using the data file of the year 1978 of the Bautzen Reservoir. The full version of the actual equation system is represented in BENNDORF (1988).

To guarantee a high transparency and flexibility of the model, the programming principles as outlined in Section 2.3 were applied. The computer program consists of four modules involving altogether 34 subroutines. There are three program versions of SALMO by using the program languages FORTRAN 77, BASIC and CSMP/360. The BASIC- and FORTRAN 77-version can be implemented on personal computers which makes possible interactive simulations.

5.2. Validation of model

The validation of the model SALMO was performed in two stages:

1. identification of uncertainties of the model by means of a comprehensive sensitivity analysis,
2. examination of the generality of the model by its application to different standing waters.

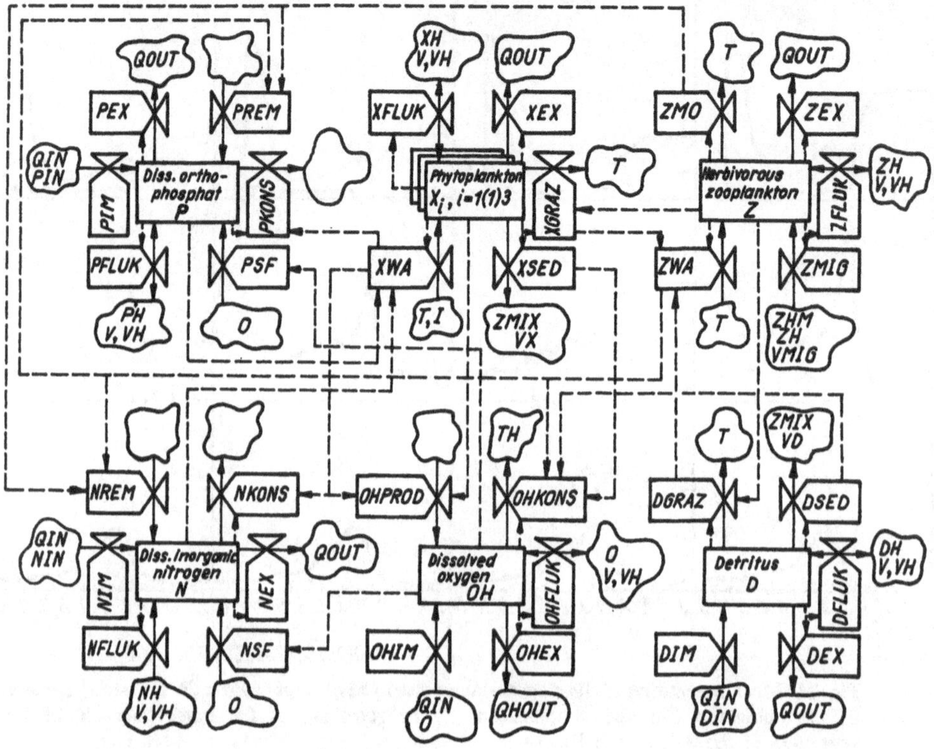

Fig. 25: Forrester-diagram of the model SALMO

Table 1: List of symbols used in the Forrester-diagram (Fig. 25) and in the documentation of state variables and process rates of the model SALMO (Fig. 26)

D	Detritus
DEX	Export of detritus by water outflow
DFLUK	Exchange of detritus between epi- and hypolimnion by fluctuating the thermocline of waters during summer
DGRAZ	Grazing of detritus by zooplankton
DH	Detritus in the hypolimnion during summer
DIM	Import of detritus by water inflow
DIN	Concentration of particular organic matter in the inflowing water
DSED	Sedimentation of detritus
I	Incident photosynthetic active radiation at the water surface
N	Dissolved inorganic nitrogen
NEX	Export of nitrogen by water outflow
NFLUK	Exchange of nitrogen between epi- and hypolimnion by fluctuating the thermocline of waters during summer
NH	Nitrogen in the hypolimnion during summer
NIM	Import of nitrogen by water inflow
NIN	Concentration of inorganic nitrogen in the inflowing water
NKONS	Nitrogen assimilation by phytoplankton
NREM	Nitrogen remineralization due to the activity of zooplankton
NSF	Nitrogen release from sediments
O	Dissolved oxygen
OH	Dissolved oxygen in the hypolimnion during summer

Table 1: (continued)

OHEX	Export of oxygen by water outflow from the hypolimnion
OHFLUK	Exchange of oxygen between the epi- and hypolimnion by fluctuating the thermocline of waters during summer
OHIM	Import of oxygen by water inflow
OHKONS	Oxygen demand in the hypolimnion except by phytoplankton respiration
OHPROD	Production and demand of oxygen by phytoplankton
P	Dissolved orthophosphate
PEX	Export of phosphate by water outflow
PFLUK	Exchange of phosphate between epi- and hypolimnion by fluctuating the thermocline of waters during summer
PH	Dissolved orthophosphate in the hypolimnion during summer
PIM	Import of phosphate by water inflow
PIN	Concentration of dissolved orthophosphate in the inflowing water
PKONS	Phosphate assimilation by phytoplankton
PREM	Phosphate remineralization due to the activity of zooplankton
PSF	Phosphate release from sediments
QIN	Water inflow
QOUT	Water outflow
T	Water temperature
TH	Water temperature in the hypolimnion
V	Volume of the water body
VH	Volume of the water in the hypolimnion
VD	Sinking velocity of detritus
VMIG	Velocity of zooplankton migration from hypo- to epilimnion during summer
VX	Sinking velocity of phytoplankton
X_i	Phytoplankton biomass of the ith functional group
XEX	Export of phytoplankton by water outflow
XFLUK	Exchange of phytoplankton between epi- and hypolimnion by fluctuating the thermocline of the waters during summer
XGRAZ	Grazing of phytoplankton by zooplankton
XH_i	Phytoplankton biomass of the ith functional group in the hypolimnion during summer
XSED	Sedimentation of phytoplankton
XWA	Growth of phytoplankton
Z	Biomass of herbivorous zooplankton
ZEX	Export of zooplankton by water outflow
ZFLUK	Exchange of zooplankton between epi- and hypolimnion by fluctuating the isocline of waters during summer
ZF	Biomass of herbivorous zooplankton in the hypolimnion during summer
ZHM	Depth of the hypolimnion
ZMIG	Migration of zooplankton from hypo- to epilimnion during summer
ZMIX	Depth of the mixed layer of waters
ZMO	Mortality of zooplankton
ZWA	Growth of zooplankton

5.2.1. Sensitivity analysis

The sensitivity analysis is generally divided into two different approaches: analytical and numerical. The numerical approach was founded by TOMOVIC and VUKOBRATOVIC in 1972 and their work initiated many papers. The analytical approach has been executed by numerous techniques. This author agrees with HALFON (1977) and SWARTZMAN (1980) that the numerical approach induces numerical errors of approximation and is

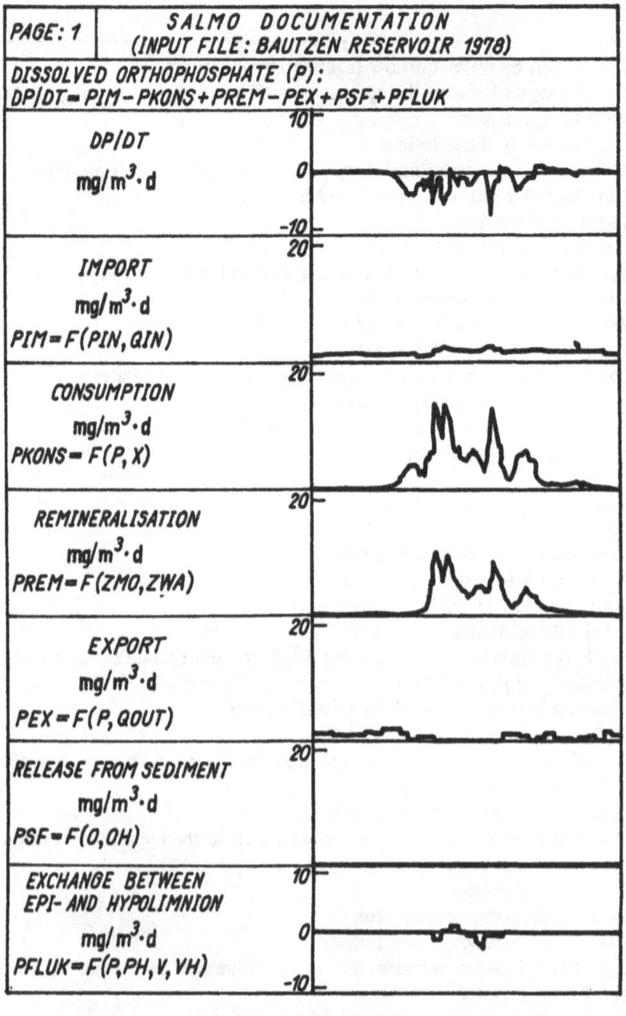

For more details of all the equations (page 1 ... page 7) see BENNDORF (1988).
Fig. 26: Documentation of the model SALMO (page 1). Dissolved orthophosphate

too complicated to make a comprehensive analysis of the complex simulations required for ecological models.

The sensitivity analysis for the model SALMO is carried out in an analytical approach using the polar coordinate technique (THORNTON et al., 1979; RECKNAGEL, 1984b). It considers: parameter sensitivity, sensitivity against the input discretization interval, initial condition sensitivity, sensitivity against randomly perturbed rate variables and numerical integration method sensitivity.

The analysis of parameter sensitivity of SALMO includes a strategy which seeks the shortest way for the highest sensitivity within the 117 parameters. The strategy is based on a hierarchical decomposition of macroparameters which are assigned endogenous system quantities: state variables, rate variables (absolute rates) and auxiliary variables

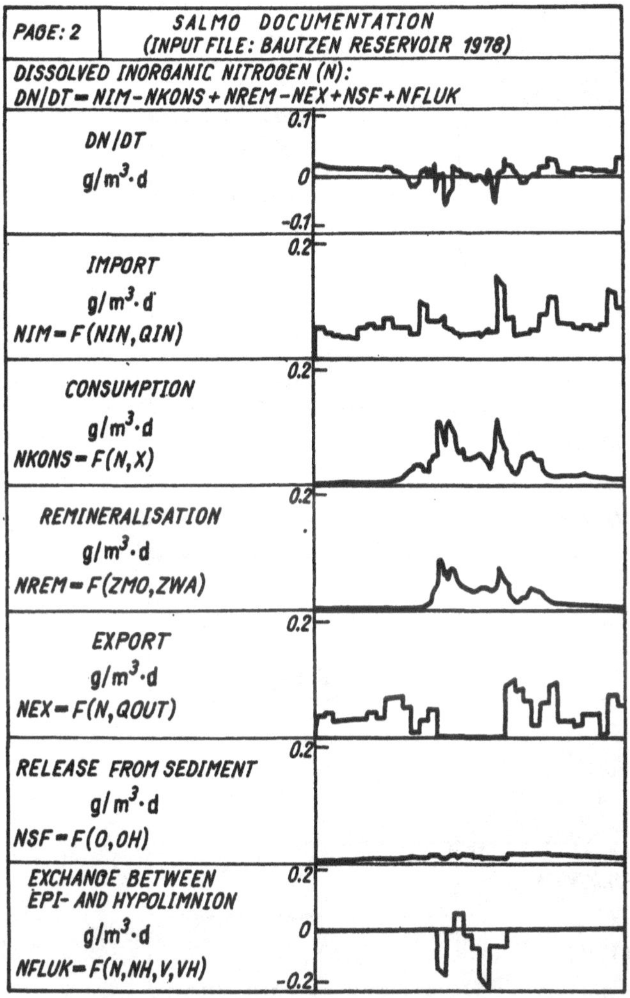

Fig. 26: Documentation of the model SALMO (page 2). Dissolved inorganic nitrogen

(including relative rates). The parameter sensitivity determined in this manner is compared with the relative variance of parameter estimates to identify directions for future research.

5.2.1.1. Method

The polar coordinate technique enables the investigator to produce compact plots of the relative model sensitivity where the day-by-day comparison of a reference standard simulation run and a perturbed simulation run for any endogenous system quantity is made (Figs. 26—38). The reference simulation run is represented as a dotted unit

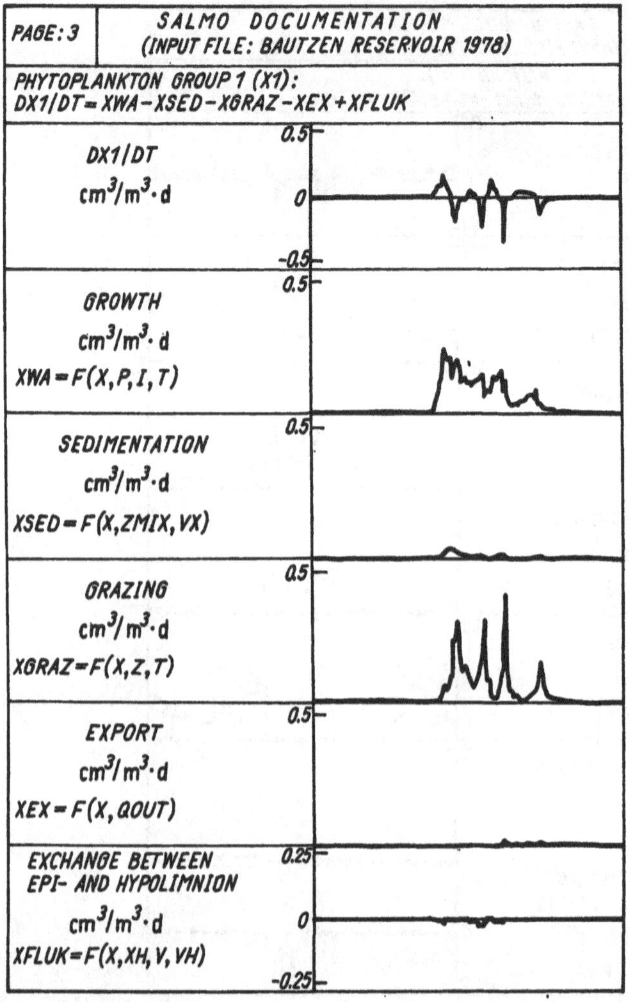

Fig. 26: Documentation of the model SALMO (page 3). Phytoplankton group 1

circle. Using 360° of the unit circle daily model sensitivity of one simulated year can be considered where the maximal sensitivity displayed by the plots amounts to 100%.

In addition to daily sensitivity, seasonal sensitivity is also represented in the plots. However, the seasonal periods of a year simulated by SALMO are individually defined for each year and are time specific for the ecosystem under consideration depending on hydrophysical conditions, namely: water body covered by a layer of ice — winter period, completely mixed water body — spring and autumn periods, stable thermic stratification of water body — summer period.

To each sensitivity plot, which refers to a specified state variable, a tabled, precise numerical sensitivity crest factor for each season is added.

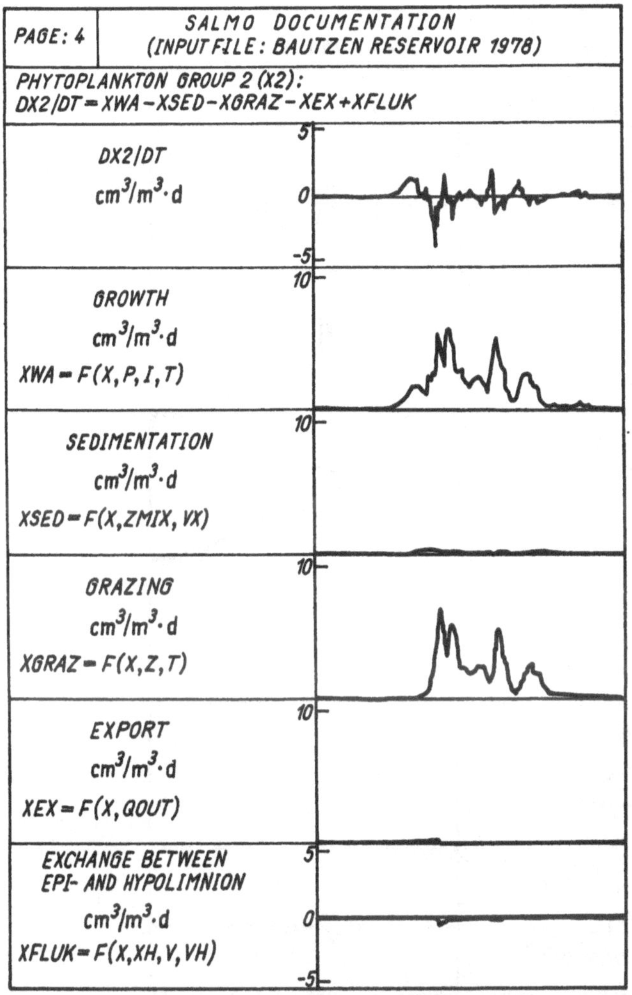

Fig. 26: Documentation of the model SALMO (page 4). Phytoplankton group 2

5.2.1.2. Parameter sensitivity

The highly complex behaviour of aquatic ecosystems requires an extensive endogenization of causal relationships of the system behaviour into ecosystem models. Such an endogenization, however, causes a disadvantageous increase of the model structure and especially of the number of parameters. SALMO seems to be a reasonable compromise between the necessary minimum of state variables and the maximal endogenization of ecosystem behaviour. Because it would be too expensive to check the model sensitivity for each of the 117 parameters the researchers have followed a seeking strategy for the highest parameter sensitivity using hierarchically decomposed macroparameters.

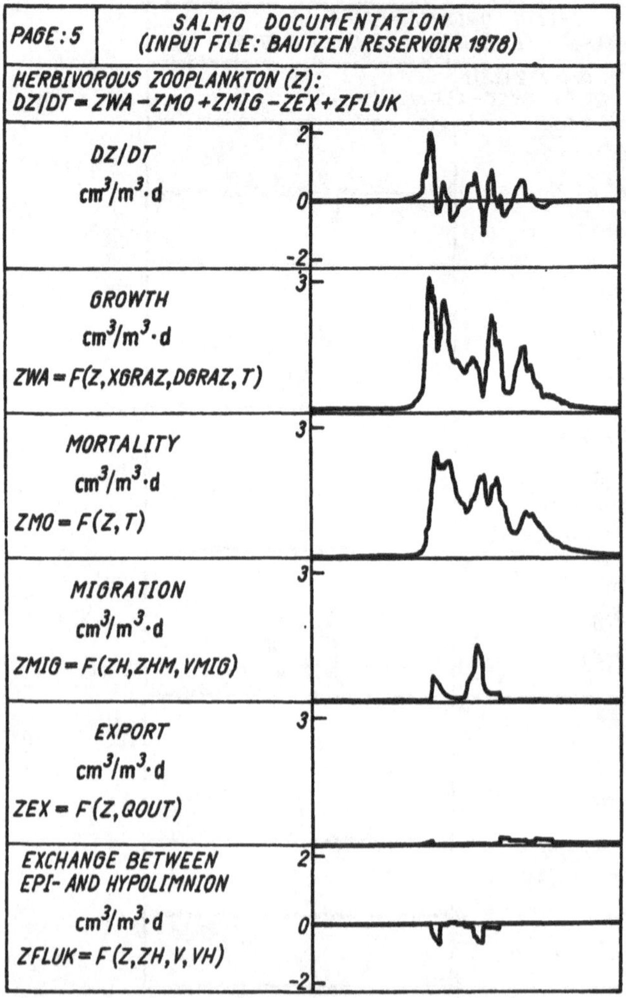

Fig. 26: Documentation of the model SALMO (page 5). Herbivorous zooplankton

Figures 27a and 27b show the state variable sensitivity with respect to a change by ±5% of the macroparameters which are assigned to the state variables. Sensitivity attained a magnitude of up to 454% of phytoplankton X1 against changing its macroparameter X1 by − 5% (Fig. 27a, left); hence, in the next step sensitivity of phytoplankton X1 against the macroparameters which are assigned the rate variables (growth, grazing by zooplankton and sedimentation of X1) was considered with the same change of −5% (Fig. 28). By reason of the dominance of sensitivity of X1 against the macroparameter referred to as the growth rate (181%; Fig. 28, left) it was split into the macroparameters of the auxiliary variables (gross-photosynthesis and respiration). As a result of checking sensitivity of X1 against a change by −5% of the macroparameters (Fig. 29) it can be seen that the influence of the respiration macroparameter clearly dominates

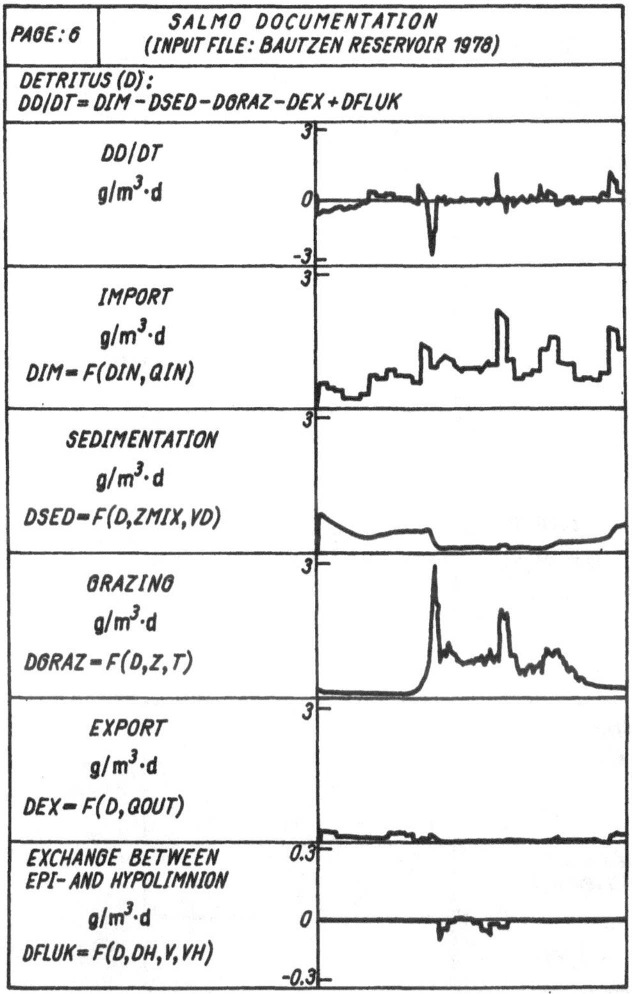

Fig. 26: Documentation of the model SALMO (page 6). Detritus

(139%; Fig. 29, right). Among the four single constant parameters of the respiration macroparameter, parameter TOPTX1 (optimum water temperature for growth of phytoplankton X1) induces the highest sensitivity in X1 at a change of −5% (40%; Fig. 30, left).

Table 2 contains the comparison between relative variance of parameter estimates and maximal parameter sensitivity related to the constant parameters TOPTX1, RXMF1 (proportion of gross-photosynthesis rate of X1 lost by respiration) and RXTMIN1 (minimal respiration rate of X1 at a temperature near 0 °C) for which the highest parameter sensitivity of phytoplankton X1 was determined (Fig. 30). TOPTX1 has a low relative variance but induces the highest parameter sensitivity of X1. The estimate of RXMF1

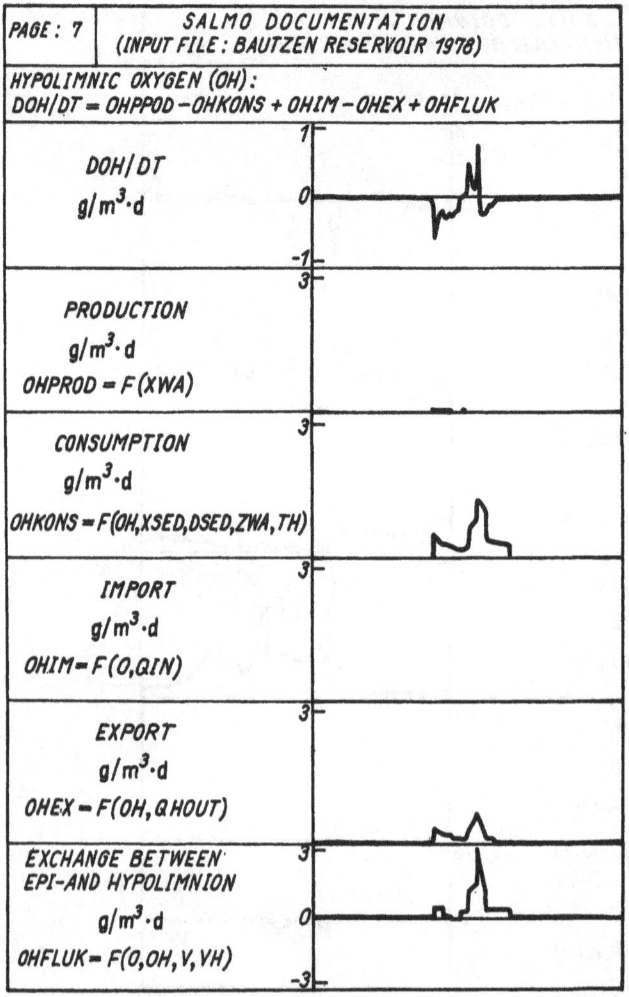

Fig. 26: Documentation of the model SALMO (page 7). Hypolimnic oxygen

exhibits the highest relative variance and causes a high parameter sensitivity of X1. Finally, relative variance of the estimated magnitude of RXTMIN1 is high but the resulting parameter sensitivity in X1 is the lowest.

Future experimental efforts should be concentrated on minimization of the variance of parameter estimates of TOPTX1 and RXMF1 by investigating them particularly in seasons of their highest parameter sensitivity on X1 (summer and autumn). Within the reduced variances it is useful to determine the new parameter estimates for the model by using an appropriate parameter optimization procedure (RECKNAGEL and BENNDORF, 1982). In this manner parameter sensitivity of each state variable of a model should be determined and interpreted to improve parameter reliability and model validity.

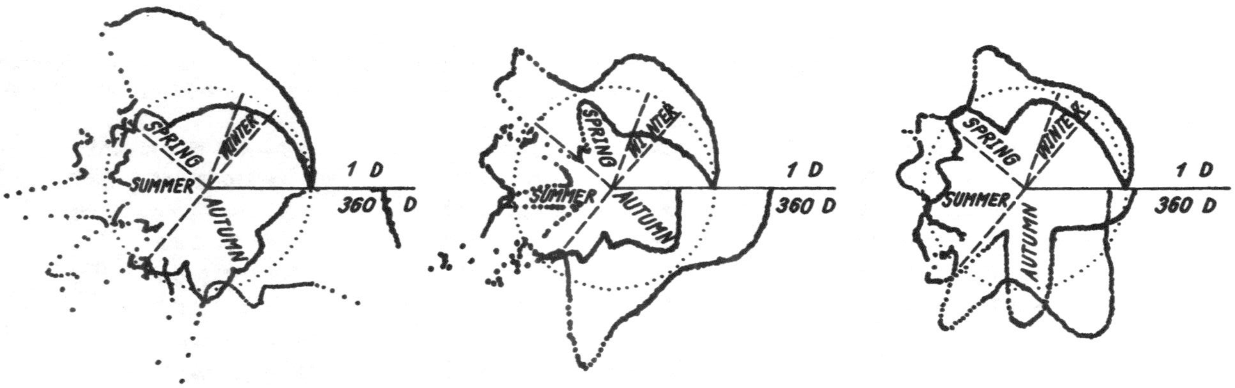

SENSITIVITY S OF : AT A CHANGE BY ±5 % OF THE MACROPARAMETER :	PHYTOPLANKTON X1 $\overline{X1}$ [1(1)34]	PHYTOPLANKTON X2 $\overline{X2}$ [1(1)34]	ZOOPLANKTON Z $\overline{Z}$ [1(1)23]
-WINTER-	S_{MIN} = 9.9 %, DAY = 50	S_{MIN} = 22.1 %, DAY = 50	S_{MIN} = 10.6 %, DAY = 69
	S_{MAX} = 32.3 %, DAY = 69	S_{MAX} = 44.5 %, DAY = 69	S_{MAX} = 14.1 %, DAY = 51
-SPRING-	S_{MIN} = 0.4 %, DAY = 135	S_{MIN} = 3.2 %, DAY = 108	S_{MIN} = 0.0 %, DAY = 133
	S_{MAX} = 103.7 %, DAY = 123	S_{MAX} = 69.3 %, DAY = 126	S_{MAX} = 46.9 %, DAY = 108
-SUMMER-	S_{MIN} = 0.3 %, DAY = 227	S_{MIN} = 0.2 %, DAY = 153	S_{MIN} = 0.3 %, DAY = 197
	S_{MAX} = 454.3 %, DAY = 220	S_{MAX} = 98.3 %, DAY = 205	S_{MAX} = 33.2 %, DAY = 225
-AUTUMN-	S_{MIN} = 0.0 %, DAY = 1	S_{MIN} = 0.0 %, DAY = 1	S_{MIN} = 0.0 %, DAY = 1
	S_{MAX} = 234.6 %, DAY = 258	S_{MAX} = 80.9 %, DAY = 262	S_{MAX} = 63.5 %, DAY = 298

Fig. 27a: Parameter sensitivity (S) of the model SALMO related to: phytoplankton biomass X1 against macroparameter $\overline{X1}$ of X1 (left); phytoplankton biomass X2 against macroparameter $\overline{X2}$ of X2 (middle); zooplankton biomass Z against macroparameter $\overline{Z}$ of Z (right)

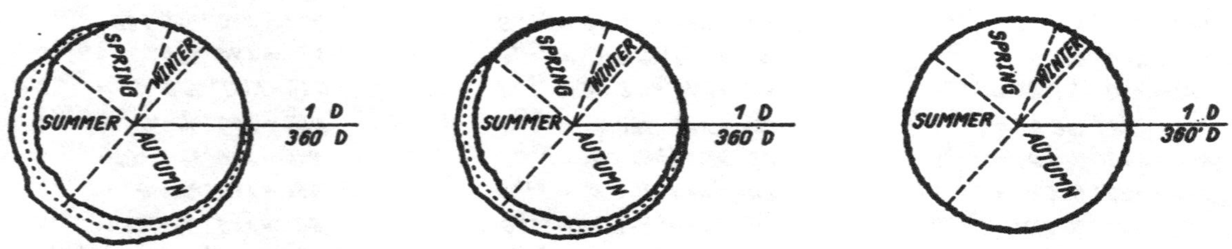

SENSITIVITY S OF : AT A CHANGE BY ±5% OF THE MACROPARAMETER :	PHOSPHATE P $\overline{P}$ [1(1)5]	NITROGEN N $\overline{N}$ [1(1)8]	OXYGEN O $\overline{O}$ [1(1)3]
-WINTER-	S_{MIN} = 0.0%, DAY = 50	S_{MIN} = 0.1%, DAY = 50	S_{MIN} = 0.0%, DAY = 50
	S_{MAX} = 0.0%, DAY = 50	S_{MAX} = 0.2%, DAY = 69	S_{MAX} = 0.1%, DAY = 60
-SPRING-	S_{MIN} = 0.0%, DAY = 70	S_{MIN} = 0.2%, DAY = 70	S_{MIN} = 0.0%, DAY = 70
	S_{MAX} = 4.6%, DAY = 123	S_{MAX} = 2.4%, DAY = 121	S_{MAX} = 0.0%, DAY = 70
-SUMMER-	S_{MIN} = 6.2%, DAY = 146	S_{MIN} = 1.9%, DAY = 140	S_{MIN} = 0.0%, DAY = 140
	S_{MAX} = 15.4%, DAY = 216	S_{MAX} = 12.1%, DAY = 212	S_{MAX} = 0.0%, DAY = 140
-AUTUMN-	S_{MIN} = 0.0%, DAY = 1	S_{MIN} = 0.0%, DAY = 1	S_{MIN} = 0.0%, DAY = 1
	S_{MAX} = 14.1%, DAY = 250	S_{MAX} = 10.2%, DAY = 251	S_{MAX} = 0.0%, DAY = 1

Fig. 27b: Parameter sensitivity (S) related to: dissolved orthophosphate P against macroparameter $\overline{P}$ of P (left); dissolved inorganic nitrogen N against macroparameter $\overline{N}$ of N (middle); dissolved oxygen O against macroparameter $\overline{O}$ of O (right)

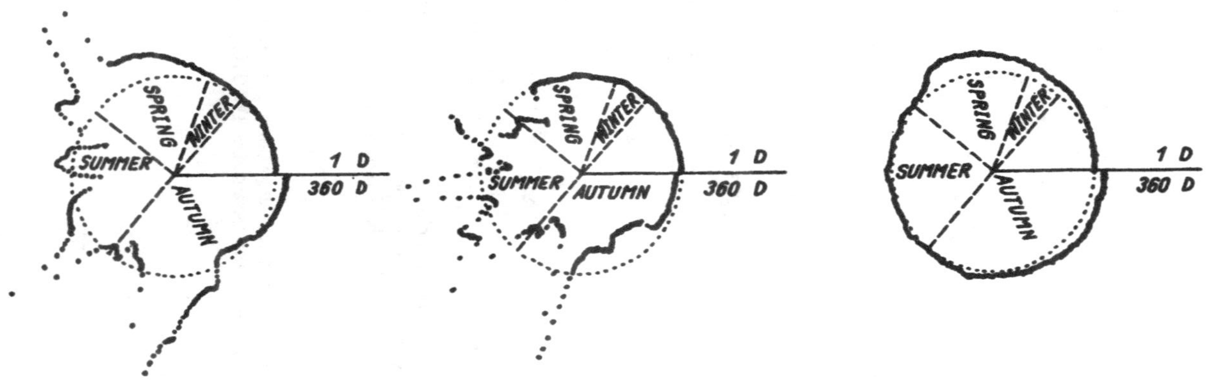

SENSITIVITY S OF: AT A CHANGE BY −5% OF THE MACROPARAMETER :	PHYTOPLANKTON X1 $\overline{GROWTH}$ [1(1)20]	PHYTOPLANKTON X1 $\overline{GRAZING}$ [1(1)13]	PHYTOPLANKTON X1 $\overline{SEDIMENTATION}$ [1(1)1]
−WINTER−	S_{MIN} = 4.0%, DAY = 50	S_{MIN} = 0.6%, DAY = 50	S_{MIN} = 9.6%, DAY = 50
	S_{MAX} = 7.3%, DAY = 69	S_{MAX} = 1.1%, DAY = 69	S_{MAX} = 13.2%, DAY = 69
−SPRING−	S_{MIN} = 7.5%, DAY = 70	S_{MIN} = 0.0%, DAY = 78	S_{MIN} = 4.5%, DAY = 139
	S_{MAX} = 112.9%, DAY = 131	S_{MAX} = 37.7%, DAY = 125	S_{MAX} = 22.6%, DAY = 116
−SUMMER−	S_{MIN} = 0.4%, DAY = 194	S_{MIN} = 0.0%, DAY = 155	S_{MIN} = 0.7%, DAY = 206
	S_{MAX} = 181.3%, DAY = 220	S_{MAX} = 136.1%, DAY = 187	S_{MAX} = 5.5%, DAY = 219
−AUTUMN	S_{MIN} = 0.0%, DAY = 1	S_{MIN} = 0.0%, DAY = 1	S_{MIN} = 1.7%, DAY = 238
	S_{MAX} = 150.3%, DAY = 258	S_{MAX} = 89.4%, DAY = 257	S_{MAX} = 11.7%, DAY = 322

Fig. 28: Parameter sensitivity (S) related to: phytoplankton biomass X1 against macroparameter $\overline{\text{GROWTH}}$ (growth rate of phytoplankton X1) (left); phytoplankton biomass X1 against macroparameter $\overline{\text{GRAZING}}$ (loss rate of phytoplankton X1 by zooplankton grazing) (middle); phytoplankton biomass X1 against macroparameter $\overline{\text{SEDIMENTATION}}$ (sedimentation rate of phytoplankton X1) (right)

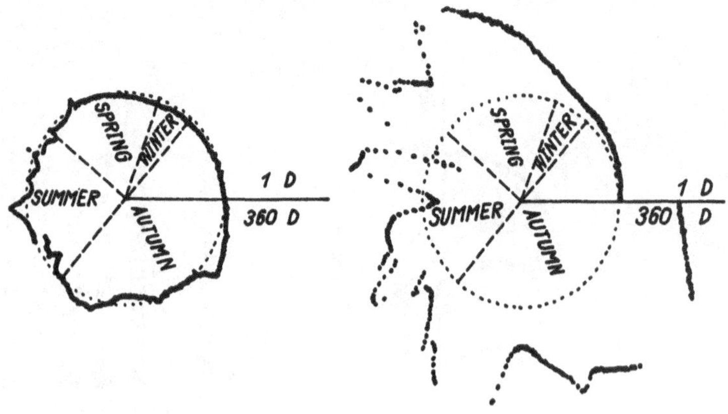

SENSITIVITY S OF AT A CHANGE BY −5 % OF THE MACROPARAMETER :	PHYTOPLANKTON X1 PHOTOSYNTHESIS [1(1)17]	PHYTOPLANKTON X1 RESPIRATION [1(1)4]
- WINTER -	S_{MIN} = 3.6 %, DAY = 50	S_{MIN} = 11.7 %, DAY = 50
	S_{MAX} = 4.5 %, DAY = 69	S_{MAX} = 22.3 %, DAY = 69
- SPRING -	S_{MIN} = 0.0 %, DAY = 101	S_{MIN} = 23.1 %, DAY = 70
	S_{MAX} = 6.8 %, DAY = 123	S_{MAX} = 106.2 %, DAY = 121
- SUMMER -	S_{MIN} = 0.1 %, DAY = 199	S_{MIN} = 0.2 %, DAY = 185
	S_{MAX} = 16.3 %, DAY = 186	S_{MAX} = 105.4 %, DAY = 148
- AUTUMN -	S_{MIN} = 0.0 %, DAY = 1	S_{MIN} = 0.0 %, DAY = 1
	S_{MAX} = 19.0 %, DAY = 323	S_{MAX} = 139.5 %, DAY = 258

Fig. 29: Parameter sensitivity (S) related to: phytoplankton biomass X1 against macroparameter PHOTOSYNTHESIS (gross-photosynthesis rate of phytoplankton X1) (left); phytoplankton biomass X1 against macroparameter RESPIRATION (respiration rate of phytoplankton X1) (right)

Tab. 2: Comparison of the relative variance of parameter estimates (Benndorf, pers. comm.) and the maximal parameter sensitivity of phytoplankton X1 related to the parameters TOPTX1, RXMF1 and RXTMIN1 (cf. Fig. 30)

Constant parameter	Relative variance of parameter estimates (%)	Maximum parameter sensitivity of X1 (%)
TOPTX1	25	39,7
RXMF1	80	21,4
RXTMIN1	70	9,1

5.2.1.3. Sensitivity against input discretization interval

There are two ways of considering driving variables in ecological models. The first one is endogenization making use of trigonometric or polynomial functions to generate smooth time series for a typical year and system. In the second way the discrete mean values

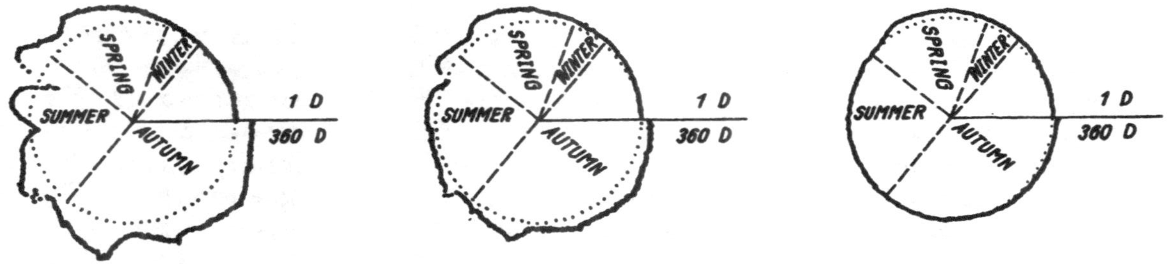

SENSITIVITY S OF: AT A CHANGE BY -5 % OF THE CONST. PARAMETER:	PHYTOPLANKTON X1 TOPTX1	PHYTOPLANKTON X1 RXMF1	PHYTOPLANKTON X1 RXTMIN1
-WINTER-	S_{MIN} = 1.5 %, DAY = 50	S_{MIN} = 3.0 %, DAY = 50	S_{MIN} = 4.6 %, DAY = 50
	S_{MAX} = 3.9 %, DAY = 69	S_{MAX} = 5.1 %, DAY = 69	S_{MAX} = 6.1 %, DAY = 69
-SPRING-	S_{MIN} = 4.1 %, DAY = 70	S_{MIN} = 5.2 %, DAY = 70	S_{MIN} = 1.2 %, DAY = 139
	S_{MAX} = 23.4 %, DAY = 121	S_{MAX} = 10.9 %, DAY = 121	S_{MAX} = 9.1 %, DAY = 114
-SUMMER-	S_{MIN} = 0.0 %, DAY = 160	S_{MIN} = 0.6 %, DAY = 159	S_{MIN} = 0.0 %, DAY = 165
	S_{MAX} = 31.9 %, DAY = 149	S_{MAX} = 17.0 %, DAY = 219	S_{MAX} = 0.9 %, DAY = 148
-AUTUMN-	S_{MIN} = 0.0 %, DAY = 1	S_{MIN} = 0.0 %, DAY = 1	S_{MIN} = 0.0 %, DAY = 1
	S_{MAX} = 39.7 %, DAY = 257	S_{MAX} = 21.4 %, DAY = 257	S_{MAX} = 4.7 %, DAY = 49

Fig. 30: Parameter sensitivity (S) related to: phytoplankton biomass X1 against constant parameter TOPTX1 (optimum water temperature for growth of phytoplankton X1) (left); phytoplankton biomass X1 against constant parameter RXMF1 (proportion of the gross-photosynthesis rate of X1 lost by respiration) (middle); phytoplankton biomass X1 against constant parameter RXTMIN1 (minimal respiration rate of X1 at a temperature near 0 °C) (right)

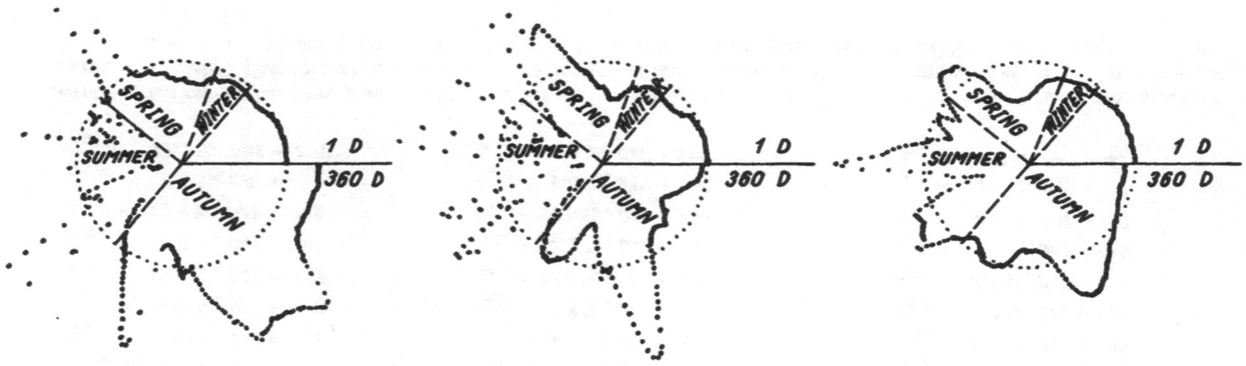

SENSITIVITY S OF: AT A CHANGE OF THE INPUT DISCRETIZATION INTERVAL FROM DECADES TO PENTADES	PHYTOPLANKTON X1	PHYTOPLANKTON X2	ZOOPLANKTON Z
-WINTER-	S_{MIN} = 5.6%, DAY = 56	S_{MIN} = 12.2 %, DAY = 61	S_{MIN} = 6.2 %, DAY = 56
	S_{MAX} = 10.5%, DAY = 137	S_{MAX} = 20.8 %, DAY = 66	S_{MAX} = 10.8 %, DAY = 70
-SPRING-	S_{MIN} = 0.5 %, DAY = 137	S_{MIN} = 0.6 %, DAY = 106	S_{MIN} = 0.7 %, DAY = 121
	S_{MAX} = 280.9 %, DAY = 126	S_{MAX} = 201.1 %, DAY = 126	S_{MAX} = 39.4 %, DAY = 105
-SUMMER-	S_{MIN} = 0.2 %, DAY = 200	S_{MIN} = 1.1 %, DAY = 178	S_{MIN} = 0.1 %, DAY = 228
	S_{MAX} = 976.2 %, DAY = 177	S_{MAX} = 137.8 %, DAY = 171	S_{MAX} = 83.5 %, DAY = 181
-AUTUMN-	S_{MIN} = 0.0 %, DAY = 1	S_{MIN} = 0.0 %, DAY = 1	S_{MIN} = 0.0 %, DAY = 1
	S_{MAX} = 94.8 %, DAY = 316	S_{MAX} = 88.4 %, DAY = 285	S_{MAX} = 42.5 %, DAY = 297

Fig. 31: Input discretization sensitivity (S) related to phytoplankton biomass X1 (left); phytoplankton biomass X2 (middle); zooplankton biomass Z (right)

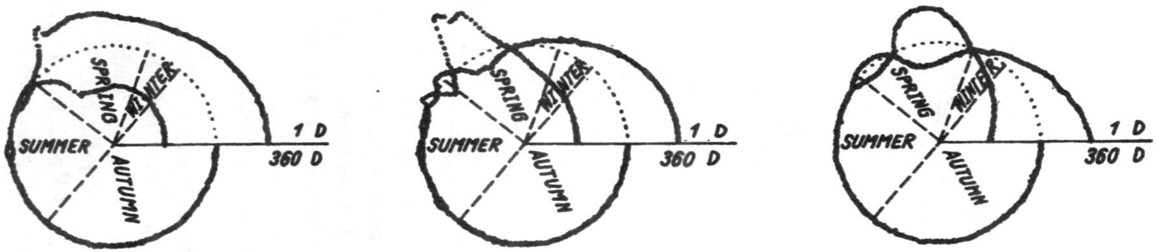

SENSITIVITY S OF : AT A CHANGE BY ±50% OF THE INITIAL CONDITION :	PHYTOPLANKTON X1	PHYTOPLANKTON X2	ZOOPLANKTON Z
	X1(0)	X2(0)	Z(0)
-WINTER-	S_{MIN}=28.4 %, DAY= 69	S_{MIN}=12.1 %, DAY= 69	S_{MIN}= 0.0 %, DAY= 68
	S_{MAX}=44.0 %, DAY= 50	S_{MAX}=40.2 %, DAY= 50	S_{MAX}=21.8 %, DAY= 50
-SPRING-	S_{MIN}= 2.6 %, DAY = 139	S_{MIN}= 0.0 %, DAY= 133	S_{MIN}= 0.0 %, DAY= 120
	S_{MAX}=40.8 %, DAY= 70	S_{MAX}= 31.9 %, DAY= 70	S_{MAX}=35.5 %, DAY= 97
-SUMMER-	S_{MIN}= 0.0 %, DAY= 166	S_{MIN}= 0.0 %, DAY= 149	S_{MIN}= 0.0 %, DAY= 156
	S_{MAX}= 5.3 %, DAY= 140	S_{MAX}= 10.1 %, DAY= 146	S_{MAX}= 6.2 %, DAY= 140
-AUTUMN-	S_{MIN}= 0.0 %, DAY= 253	S_{MIN}= 0.0 %, DAY= 243	S_{MIN}= 0.0 %, DAY= 238
	S_{MAX}= 50.4 %, DAY= 7	S_{MAX}=50.1 %, DAY= 3	S_{MAX}= 49.8 %, DAY= 1

Fig. 32: Initial condition sensitivity (S) related to: phytoplankton biomass X1 against its initial value X1(0) (left); phytoplankton biomass X2 against its initial value X2(0) (middle); zooplankton biomass Z against its initial value Z(0) (right)

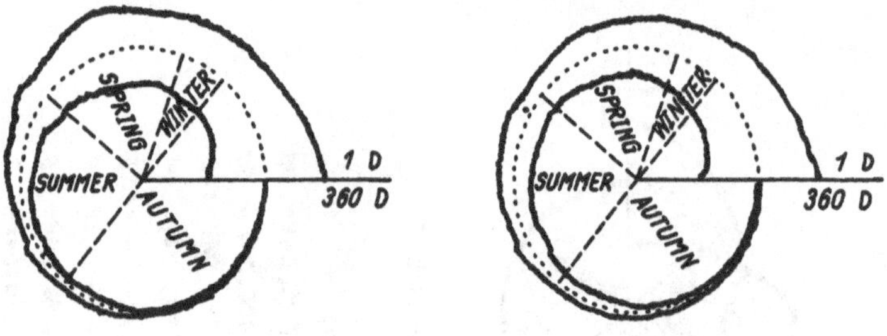

SENSITIVITY S OF: AT A CHANGE BY ±50% OF THE INITIAL CONDITION:	PHOSPHATE P P(0)	NITROGEN N N(0)
-WINTER-	S_{MIN} = 29.7%, DAY = 69	S_{MIN} = 25.0%, DAY = 69
	S_{MAX} = 32.0%, DAY = 50	S_{MAX} = 31.6%, DAY = 50
-SPRING-	S_{MIN} = 18.0%, DAY = 139	S_{MIN} = 12.6%, DAY = 139
	S_{MAX} = 29.6%, DAY = 70	S_{MAX} = 25.0%, DAY = 70
-SUMMER-	S_{MIN} = 4.7%, DAY = 225	S_{MIN} = 7.9%, DAY = 225
	S_{MAX} = 21.9%, DAY = 146	S_{MAX} = 12.5%, DAY = 140
-AUTUMN-	S_{MIN} = 0.6%, DAY = 360	S_{MIN} = 1.2%, DAY = 360
	S_{MAX} = 50.0%, DAY = 1	S_{MAX} = 50.0%, DAY = 1

Fig. 33: Initial condition sensitivity (S) related to: dissolved orthophosphate P against its initial value P (0) (left); inorganic dissolved nitrogen N against its initial value N (0) (right)

of actual measurements are used to observe the model's response to actual crest factors in hydrological, meteorological and physical conditions of a year under consideration. The author favours the second way in order to fit the model for scenario analysis. Since data winning by routine measurements and data processing is time consuming, for this way there should be found a compromise between the required length of the discretization interval and model validity.

Figure 31 shows that the shortening of the discretization interval for the whole input file from decades to pentades effects a high model sensitivity: however, subsequent investigations have shown that only hydrophysical input variables, such as ZMIXREAL (real mixing depth) and ZMIX (mean mixing depth), are responsible (HEERKLOTZ, 1982). The discretization interval of the input variables ZMIXREAL and ZMIX, therefore, should be chosen as short as possible for careful determination. For the residual 21 input variables it is acceptable to use decades as their discretization interval.

5.2.1.4. Initial condition sensitivity

Checking the sensitivity of an ecological model against the initial conditions at a change of ±50% reveals the resilience of its endogenous dynamics.

Figure 32 shows that a high sensitivity, about 40%, of biological state variables X1, X2 and Z against a change of 50% in their initial conditions is spreading up to the beginning of the summer period but then is easing down rapidly towards 0%. The initial condition sensitivity of the nutrient state variables P and N decreases asymptotically to almost 0% in the course of the year (Fig. 33).

This study shows that SALMO exhibits a stable annual rhythm because its state variables remain steady under such strongly changed initial conditions.

5.2.1.5. Sensitivity against randomly perturbed rate variables

Checking the model sensitivity against randomly perturbed rate variables displays the resistance of its endogenous dynamics.

In this experiment the essential rate variables (r) for the determination of biological state variables X1, X2 and Z are randomly perturbed by $\pm 50\%$ after daily calculations using a pseudorandom digit (prd) as follows:

$$r = r/2 + \text{prd} \cdot r \qquad (0 \leq \text{prd} \leq 1) .$$

Figure 34 shows the phytoplankton X1 response to be highly sensitive (more than 100%) to the $\pm 50\%$ perturbation of its rate variables ‚growth‘ and ‚grazing‘ in summer and autumn. Nevertheless, perturbed trajectories return to the reference value of the standard run at the end of the year rather than becoming out of control. The influence of ‚sedimentation‘ perturbation (Fig. 34, right) is relatively low but also noticeable in summer and autumn.

The same assessment applies to phytoplankton X2 behaviour against $\pm 50\%$ perturbation of growth, grazing and sedimentation rates (Fig. 35) where a high sensitivity (about 80%) of X2 to perturbed growth rate (Fig. 35, left) is present already in spring.

In Fig. 36 the sensitivity of zooplankton Z against randomly perturbed rate variables ‚growth‘ and ‚mortality‘ by $\pm 50\%$ is depicted as being highest against growth in spring (54%; Fig. 36, left) and against mortality in autumn (33%; Fig. 36, right). In both cases perturbed trajectories nearly return to the reference value of the standard run at the end of the year.

Sensitivity analysis against randomly perturbed rate variables of biological state variables has shown that SALMO involves a robustly endogenous dynamics by the functioning of feedback control mechanisms.

5.2.1.6. Numerical integration method sensitivity

Determining the numerical integration method sensitivity enables the investigator to evaluate the fitness of the selected integration method for standard simulation runs with respect to the numerical stability of a differential equations solution.

Use of the CSMP/360-version of SALMO has allowed the generation of simulation runs with six available methods of CSMP: rectangular method, trapezoidal method, Adam method, Simpson method, Milne method and Runge-Kutta method. In the standard run of SALMO differential equations are solved by the four-step Runge-Kutta method with fixed intervals. Subsequently each of the other five methods of CSMP is compared with the reference run related to solving the differential equation of phytoplankton biomass X1 as a case study (Figs. 37 and 38). The highest sensitivity of Runge-

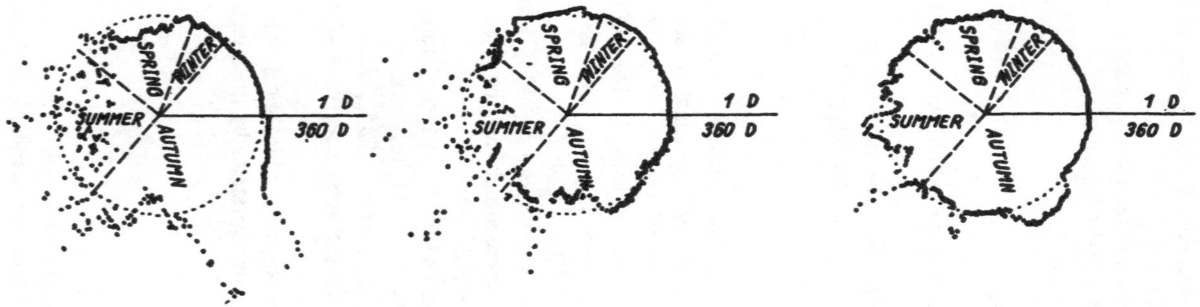

SENSITIVITY S OF AT A PSEUDO-RANDOM PERTURBATION BY ±50% OF THE RATE VARIABLE	PHYTOPLANKTON X1 GROWTH	PHYTOPLANKTON X1 GRAZING	PHYTOPLANKTON X1 SEDIMENTATION
-WINTER-	S_{MIN} = 0.1%, DAY = 64	S_{MIN} = 4.9%, DAY = 61	S_{MIN} = 1.4%, DAY = 61
	S_{MAX} = 5.1%, DAY = 68	S_{MAX} = 8.2%, DAY = 51	S_{MAX} = 4.4%, DAY = 51
-SPRING-	S_{MIN} = 0.3%, DAY = 75	S_{MIN} = 0.6%, DAY = 134	S_{MIN} = 0.0%, DAY = 133
	S_{MAX} = 42.7%, DAY = 138	S_{MAX} = 17.9%, DAY = 127	S_{MAX} = 23.0%, DAY = 123
-SUMMER-	S_{MIN} = 0.1%, DAY = 169	S_{MIN} = 1.0%, DAY = 154	S_{MIN} = 0.0%, DAY = 196
	S_{MAX} = 73.0%, DAY = 160	S_{MAX} = 170.5%, DAY = 192	S_{MAX} = 58.9%, DAY = 219
-AUTUMN-	S_{MIN} = 0.0%, DAY = 33	S_{MIN} = 0.1%, DAY = 334	S_{MIN} = 0.0%, DAY = 278
	S_{MAX} = 127.1%, DAY = 305	S_{MAX} = 63.5%, DAY = 258	S_{MAX} = 26.8%, DAY = 257

Fig. 34: Sensitivity (S) against randomly perturbed rate variables related to: phytoplankton biomass X1 against rate variable GROWTH (growth rate of X1) (left); phytoplankton biomass X1 against rate variable GRAZING (loss rate of X1 by zooplankton grazing) (middle); phytoplankton biomass X1 against rate variable SEDIMENTATION (sedimentation rate of X1) (right)

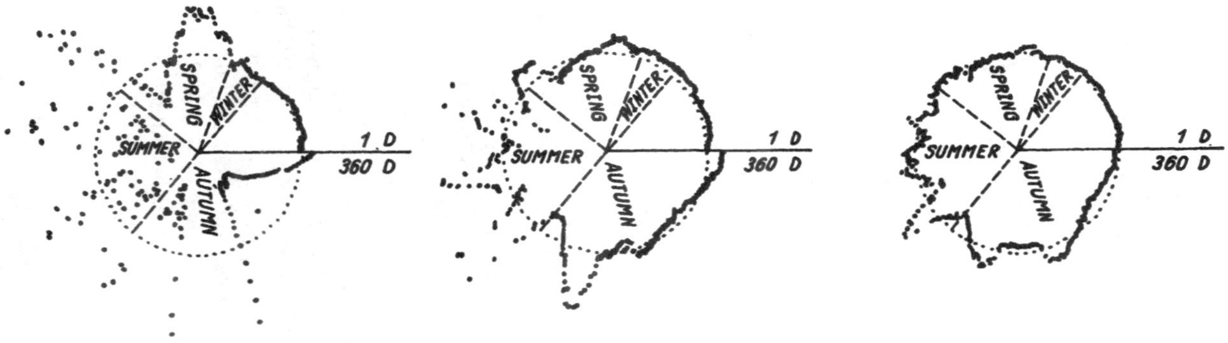

SENSITIVITY S OF AT A PSEUDO-RANDOM PERTURBATION BY ±50% OF THE RATE VARIABLE :	: PHYTOPLANKTON X2 GROWTH	PHYTOPLANKTON X2 GRAZING	PHYTOPLANKTON X2 SEDIMENTATION
-WINTER-	S_{MIN} = 0.4 %, DAY = 65	S_{MIN} = 7.8 %, DAY = 57	S_{MIN} = 2.6 %, DAY = 61
	S_{MAX} = 6.8 %, DAY = 68	S_{MAX} = 11.4 %, DAY = 65	S_{MAX} = 5.9 %, DAY = 51
-SPRING-	S_{MIN} = 0.7 %, DAY = 72	S_{MIN} = 0.2 %, DAY = 103	S_{MIN} = 0.0 %, DAY = 92
	S_{MAX} = 79.4 %, DAY = 138	S_{MAX} = 19.2 %, DAY = 139	S_{MAX} = 12.8 %, DAY = 123
-SUMMER-	S_{MIN} = 0.4 %, DAY = 212	S_{MIN} = 0.1 %, DAY = 209	S_{MIN} = 0.6 %, DAY = 159
	S_{MAX} = 345.8 %, DAY = 230	S_{MAX} = 204.0 %, DAY = 219	S_{MAX} = 39.7 %, DAY = 218
-AUTUMN-	S_{MIN} = 0.0 %, DAY = 42	S_{MIN} = 0.0 %, DAY = 271	S_{MIN} = 0.0 %, DAY = 308
	S_{MAX} = 290.9 %, DAY = 281	S_{MAX} = 83.5 %, DAY = 258	S_{MAX} = 18.9 %, DAY = 254

Fig. 35: Sensitivity (S) against randomly perturbed rate variables related to: phytoplankton biomass X2 against rate variable GROWTH (growth rate of X2) (left); phytoplankton biomass X2 against rate variable GRAZING (loss rate of X2 by zooplankton grazing) (middle); phytoplankton biomass X2 against rate variable SEDIMENTATION (sedimentation rate of X2) (right)

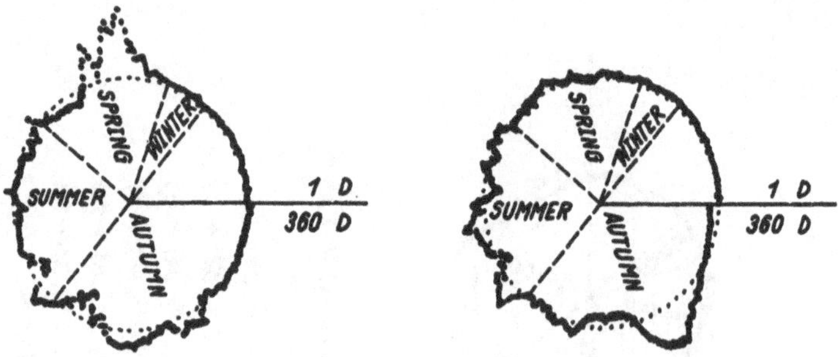

SENSITIVITY S OF: AT A PSEUDO-RANDOM PERTURBATION BY ±50% OF THE RATE VARIABLE:	: ZOOPLANKTON Z GROWTH	ZOOPLANKTON Z MORTALITY
-WINTER-	S_{MIN} = 0.9 %, DAY = 51	S_{MIN} = 2.0 %, DAY = 64
	S_{MAX} = 1.5 %, DAY = 57	S_{MAX} = 4.2 %, DAY = 51
-SPRING-	S_{MIN} = 0.2 %, DAY = 115	S_{MIN} = 0.2 %, DAY = 102
	S_{MAX} = 54.0 %, DAY = 96	S_{MAX} = 6.3 %, DAY = 119
-SUMMER-	S_{MIN} = 0.1 %, DAY = 205	S_{MIN} = 0.1 %, DAY = 157
	S_{MAX} = 15.3 %, DAY = 214	S_{MAX} = 20.6 %, DAY = 224
-AUTUMN-	S_{MIN} = 0.0 %, DAY = 312	S_{MIN} = 0.0 %, DAY = 236
	S_{MAX} = 24.1 %, DAY = 247	S_{MAX} = 32.6 %, DAY = 304

Fig. 36: Sensitivity (S) against randomly perturbed rate variables related to: zooplankton biomass Z against rate variable GROWTH (growth rate of Z) (left); zooplankton biomass Z against rate variable MORTALITY (mortality rate of Z) (right)

Kutta method was found, against the rectangular and Adam's methods, to be about 20% (Fig. 37, left and right). The lowest sensitivity was found against Simpson and Milne methods with about 9% (Fig. 38). By reason of the relatively low sensitivity against Milne's highly sophisticated but time-consuming five-step predictor-corrector method with variable intervals taking into account the numerical stability, the author accepted the precision of Runge-Kutta method for the standard run of SALMO. At the same time the numerical integration method was excluded as a weighty source of uncertainties during the validation of SALMO.

5.2.1.7. Conclusions

Analytical sensitivity analysis, using the polar coordinate method, proved to be suitable for a comprehensive sensitivity analysis of an ecological model.

Parameter sensitivity analysis for ecological models with a high number of parameters requires an appropriate seeking strategy for the highest parameter sensitivity of each state variable. Using macroparameters related to endogenous system quantities of the model makes a hierarchical shortest-way strategy possible.

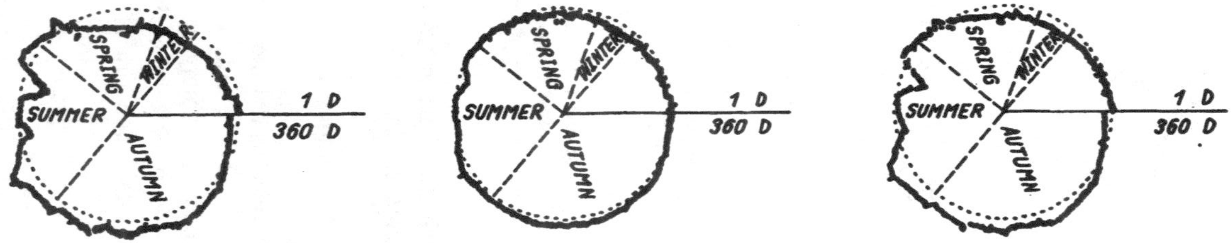

SENSITIVITY S OF RUNGE-KUTTA'S METHOD FOR NUMERICAL INTEGRATION OF : AGAINST :	PHYTOPLANKTON X1 RECTANGULAR METHOD	PHYTOPLANKTON X1 TRAPEZOIDAL METHOD	PHYTOPLANKTON X1 ADAM'S METHOD
-WINTER-	S_{MIN} = 13.9 %, DAY = 69	S_{MIN} = 5.1 %, DAY = 69	S_{MIN} = 7.3 %, DAY = 51
	S_{MAX} = 15.1 %, DAY = 56	S_{MAX} = 6.3 %, DAY = 62	S_{MAX} = 10.5 %, DAY = 54
-SPRING-	S_{MIN} = 1.5 %, DAY = 125	S_{MIN} = 0.3 %, DAY = 121	S_{MIN} = 0.3 %, DAY = 133
	S_{MAX} = 23.3 %, DAY = 92	S_{MAX} = 11.6 %, DAY = 92	S_{MAX} = 16.1 %, DAY = 92
-SUMMER-	S_{MIN} = 0.0 %, DAY = 172	S_{MIN} = 0.0 %, DAY = 154	S_{MIN} = 0.1 %, DAY = 195
	S_{MAX} = 22.1 %, DAY = 212	S_{MAX} = 7.4 %, DAY = 161	S_{MAX} = 25.1 %, DAY = 162
-AUTUMN-	S_{MIN} = 0.0 %, DAY = 333	S_{MIN} = 0.0 %, DAY = 342	S_{MIN} = 0.0 %, DAY = 334
	S_{MAX} = 15.2 %, DAY = 232	S_{MAX} = 7.5 %, DAY = 43	S_{MAX} = 13.8 %, DAY = 272

Fig. 37: Numerical integration method sensitivity (S) related to: solution of the differential equation of phytoplankton biomass X1 by Runge-Kutta's method against the rectangular method (left); solution of the differential equation of phytoplankton biomass X1 by Runge-Kutta's method against the trapezoidal method (middle); solution of the differential equation of phytoplankton biomass X1 by Runge-Kutta's method against Adam's method (right)

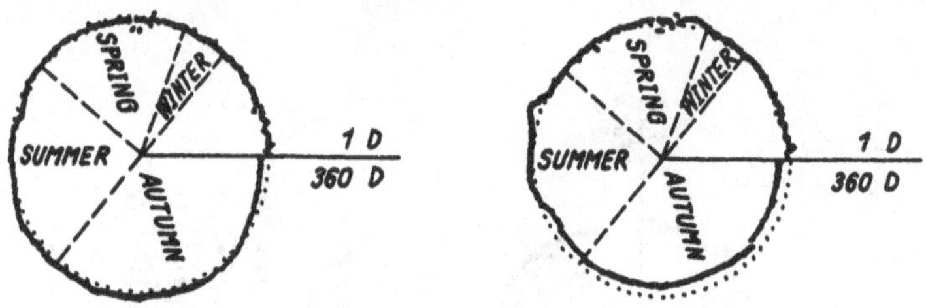

SENSITIVITY S OF RUNGE-KUTTA'S METHOD FOR NUMERICAL INTEGRATION OF: AGAINST:	PHYTOPLANKTON X1 SIMPSON'S METHOD	PHYTOPLANKTON X1 MILNE'S METHOD
-WINTER-	S_{MIN} = 1.2 %, DAY = 69	S_{MIN} = 0.9 %, DAY = 56
	S_{MAX} = 1.9 %, DAY = 50	S_{MAX} = 2.1 %, DAY = 69
-SPRING-	S_{MIN} = 0.0 %, DAY = 121	S_{MIN} = 0.0 %, DAY = 121
	S_{MAX} = 9.5 %, DAY = 92	S_{MAX} = 6.6 %, DAY = 84
-SUMMER-	S_{MIN} = 0.0 %, DAY = 153	S_{MIN} = 0.2 %, DAY = 175
	S_{MAX} = 5.5 %, DAY = 228	S_{MAX} = 8.4 %, DAY = 211
-AUTUMN-	S_{MIN} = 0.0 %, DAY = 26	S_{MIN} = 0.0 %, DAY = 13
	S_{MAX} = 6.2 %, DAY = 356	S_{MAX} = 9.2 %, DAY = 293

Fig. 38: Numerical integration method sensitivity (S) related to: solution of the differential equation of phytoplankton biomass X1 by Runge-Kutta's method against Simpson's method (left); solution of the differential equation of phytoplankton biomass X1 by Runge-Kutta's method against Milne's method (right)

Comparison between parameter sensitivity and relative variance of parameter estimates indicates directions for future experimental efforts to increase parameter reliability and model validity.

The input discretization sensitivity is worth knowing to find a compromise between the expense for data winning, or processing, and model validity.

The initial condition sensitivity reveals the resilience of the model's endogenous dynamics while sensitivity against randomly perturbed rate variables reveals the resistance of the model's endogenous dynamics as relative measures of behavioural stability of a model. Sensitivity analysis related to numerical integration methods is able to evaluate the precision of differential equation solutions of the standard simulation run of a model with respect to numerical stability.

5.2.2. Application of the model to different waters

To judge generality of the model it was applied to a multitude of different waters where comparisons of measured and simulated trajectories were taken as a basis. These comparisons for six standing waters of different trophic levels and different types of uses are illustrated in Figs. 39 and 40. The state trajectories of orthophosphate, phytoplankton (total biomass) and zooplankton have served as criteria of comparison. Uniform parameter values were used without calibrating the model.

In Fig. 39, the oligotrophic Lake Stechlin, the mesotrophic Saidenbach Reservoir and

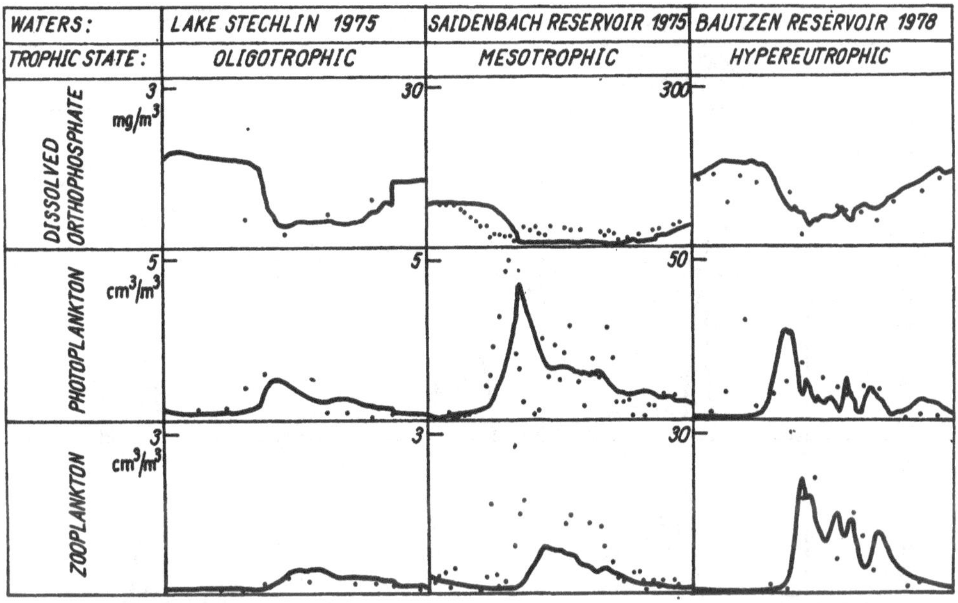

Fig. 39: Validation of the model SALMO by the hand of measured and simulated state trajectories of orthophosphate, phytoplankton (total biomass) and zooplankton for waters of different trophic levels:
— Lake Stechlin 1975 (measuring data after Benndorf, Koschel and Recknagel (1985))
— Saidenbach Reservoir 1975 (measuring data after Horn (1978))
— Bautzen Reservoir 1978 (measuring data after Benndorf (1979b)) (See also Benndorf and Recknagel (1982), Recknagel and Benndorf (1982))

the hypereutrophic Bautzen Reservoir are subjects of investigation. With regard to seasonal periodicity, qualitative conformity of measured and simulated trajectories of these waters is generally assessed as being satisfactory. However, the detailed quantitative consideration to some extent reveals considerable deviations especially concerning spring time aspects of phytoplankton and, consequently, also dynamics of orthophosphate and of zooplankton. These deviations primarily occur as a time delay of the simulated trajectories as against the measured trajectories of phytoplankton. The delay is likewise observed for the different waters where its length amounts to three to four decades. The maximum length of delay comprising four decades occurs for the hypereutrophic Bautzen Reservoir (Fig. 39) and for the potentially high-eutrophic Bleiloch Reservoir (Fig. 40). The hypothesis by Benndorf (1980) already mentioned must be regarded as a possible reason for the fact that by a high photosynthesis a minimization of the phytoplankton respiration is involved. Obviously, the intensity of this process increases with increasing eutrophic level. Apart from data inaccuracies always to be reckoned with by this unrealistic course of the trajectories in spring time, deviations during the summer months are explained as well.

More detailed discussions of simulation results are undertaken by Recknagel and Benndorf (1982) referred to the Saidenbach Reservoir, by Benndorf and Recknagel (1982) referred to the Bautzen and Bleiloch Reservoirs and by Benndorf, Koschel and Recknagel (1985) referred to the Lake Stechlin.

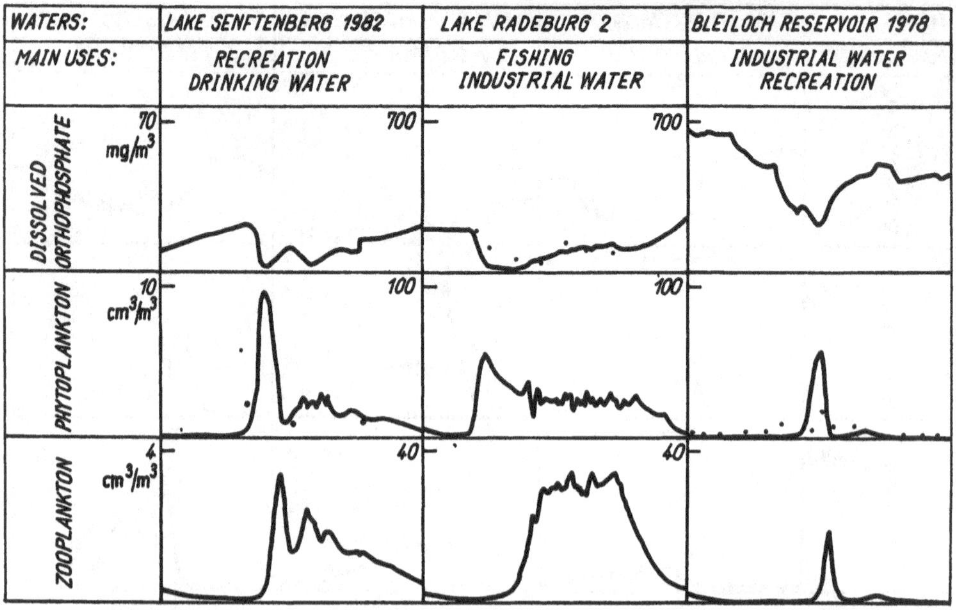

Fig. 40: Validation of the model SALMO by the hand of measured and simulated state trajectories of orthophosphate, phytoplankton (total biomass) and zooplankton for waters of different main uses:
— Lake Senftenberg 1982 (measuring data after PÜTZ and KNESCHKE, pers. comm.)
— Lake Radeburg 2 (measuring data after PÜTZ and KNESCHKE, pers. comm.)
— Bleiloch Reservoir 1978 (measuring data after LOTH, pers. comm.) (See also BENNDORF and RECKNAGEL (1982))

It is to be demonstrated by Fig. 40 that the model supplies realistic simulation results for waters of different main uses, too. However, the comparison between state trajectories measured and state trajectories simulated is only based on some few measuring data.

The Lake Senftenberg and the Lake Radeburg 2 represent standing waters underlying unusual conditions. Whereas the Lake Senftenberg originates from the seventies only within a recultivation of an reconditioned open mine and, thus, is still a „young" anthropogenously created water, the Lake Radeburg 2 was used as a carp pond in the sense of a production ecosystem during simulation. The state of the Bleiloch Reservoir at the time of simulation is assessed as being extreme because different types of loadings of the drainage area overlapped and resulted in atypical seasonal variations regarding phytoplankton and zooplankton. In addition to a high phosphorus load of about 16.5 mg P/m^2 y, the Bleiloch Reservoir was also subject to a considerable brown coloration by waste waters from pulp industry resulting in a high light extinction and thus the light limitation of phytoplankton growth. Only at a low layer thickness of the epilimnion at the beginning of summer stratification, for a short time a pulse-like phytoplankton peak and, subsequently, a zooplankton peak appeared. Regardless of the temporal delaying effects discussed above, these events are realistically simulated by the model.

In addition to the discussed criteria of trophic level and water uses to evaluate generality of the model SALMO, Figs. 39 and 40 additionally imply morphometric criteria. The waters considered cover a mean water depth of 2.5 m (Lake Radeburg 2) up to 23 m (Lake Stechlin) and water surface of $146 \cdot 10^4$ m^2 (Saidenbach Reservoir) up to approximately 9 km^2 (Lake Senftenberg).

By examining generality of the model SALMO it is concluded that the model generates a realistic system behaviour and, consequently, possesses a predictive validity for a wide range class of standing waters defined according to criteria of trophy, morphometry and uses. In future, much attention is to be paid to improve the quantitative accuracy by causal relations.

6. Model-based Water Quality Control of Lakes and Reservoirs

Management control or strategic planning, respectively, of lakes and reservoirs are aimed at minimizing the expenses for the treatment of raw water to make it available for relevant uses (see Fig. 2 in Chapter 1). To meet this target policies have to be found which prevent eutrophication and saprobization processes in an optimal manner. As a starting point for policy making in the context of this chapter the GDR-Standard Utilization and Protection of Waters (TGL 27885/03) which includes the set of feasible measures for water quality control serves. In correspondence with Section 2.4, Figures 13 and 15c the measures can be distinguished into three classes with different preferences: external preventive measures according to the open-loop control of the system environment, internal preventive control according to the closed-loop control with feedback and feed-forward, restauration measures according to the closed-loop control with feedback. In Tab. 3 the three classes of control measures are summarized.

In Fig. 41 the off-line approach of the model-based water quality control of optional lakes or reservoirs is represented. On the understanding of measured input vector ***X*** and output vector ***Y*** the optimum control trajectories of vector ***U*** can be determined where in the framework of computer simulation different methods of systems analysis are included in an integrated manner. Computer simulation is performed by means of the model SALMO represented in chapter 5.

An *expertise* is made by including regional water quality authorities to define the nominal vector ***U*** of control measures with respect to technological, institutional and economic constraints of the specific water under consideration. On the basis of the nominal vector ***U*** *scenario analysis* permits to identify alternative control policies by investigating response behaviour of the model in the „what if"-mode. However, the results of scenario analysis refer to a restricted system behaviour because only point estimations, by holding control quantities constant, can be used. To overcome this restriction, *stability analysis* by using the catastrophe theoretical concept (Recknagel, 1985) is involved for investigating the systems transient behaviour within the whole control space under consideration. In this manner stable regions of alternative control spaces can be identified as criteria to assess the preference of underlying policies. *Dynamic cost optimization* with state space constraints (Recknagel, 1984a) is directed to an optimal control of the water in regard of costs as well as water quality. In this way the most preferable policy can be determined.

Finally, a *risk analysis* has to be performed to evaluate the reliability of determined policies and control trajectories under influence of potential risk factors like uncertainties in decision making. The risk can be quantified by the probability density function of predicted variables with respect to specific sources of uncertainties. According to Suter, Barnthouse

Table 3: Measures for eutrophication control in lakes and reservoirs according to the GDR-Standard Utilization and Protection of Waters

Measures for external preventive control:
- Introduction of water-saving technologies in industrial establishments
- Diversion of sewage from the catchment area
- Evacuation or restauration of live-stock farms
- Restrictions to fertilization of agricultural areas
- Construction of sewage treatment plant including tertiary treatment (P-elimination)
- Construction of pre-dams and plant basins for biological P-elimination
- P-elimination plant for flocculation and filtration of inflowing water (Wahnbach-method)

Measures for internal preventive control:
- Manipulation of food-web by high stocks of predatory fish
- Control of bioproduction by artificial feeding of zooplankton during winter
- Dilution or flushing of waters by water poor in nutrients
- Control of bacterial N-elimination
- Deep water release
- Chemical measures for neutralization of the waters
- Chemical measures for precipitation of nutrients and co-precipitation of phytoplankton into the sediment of waters
- Control of internal P-load from sediment by lake bottom sealing
- Artificial destratification of water body during summer
- Artificial aeration of hypolimnion during summer
- Artificial light screening of water for inhibition of primary production
- Increase of the mean depth of waters

Restauration measures:
- Sediment dredging
- Chemical measures for the control of undesirable vegetables in waters

and O'NEILL (1987) sources of uncertainty in model-based decision making for environmental systems can be distinguished as follows:

(1) Model construction errors such as: aggregation errors, choosing incorrect functions for interactions between variables or setting inappropriate boundaries for components of the real system to be included in the model.

(2) Natural stochasticity of environmental systems caused by spatial heterogeneity, temporal variability and genetic heterogeneity.

(3) Errors in measurements of ecological variables and parameter estimates.

Different techniques can be used to quantify probability density functions for predicted variables of ecological models. After BECK (1987) for analysis of uncertainty in the area of water quality control following approaches promise to be convenient:

(1) Monte Carlo simulation to calculate distributions of water quality characteristics that would result from alternative management scenarios or to assess the risk of violating a prescribed standard (e.g. O'NEILL et al. (1982), FEDRA (1983), FONTAINE and LESHT (1987))

(2) Constrained optimization under uncertainty

(3) Extension of method (2) by fuzzy logic in order to derive „satisfying“, as opposed to optimizing, solutions

(4) Application the fuzzy logic in the field of knowledge engineering as an operational means of gauging the performance of a pollution control authority or as a means of a more qualitative system simulation and prediction in an interdisciplinary, decision environment.

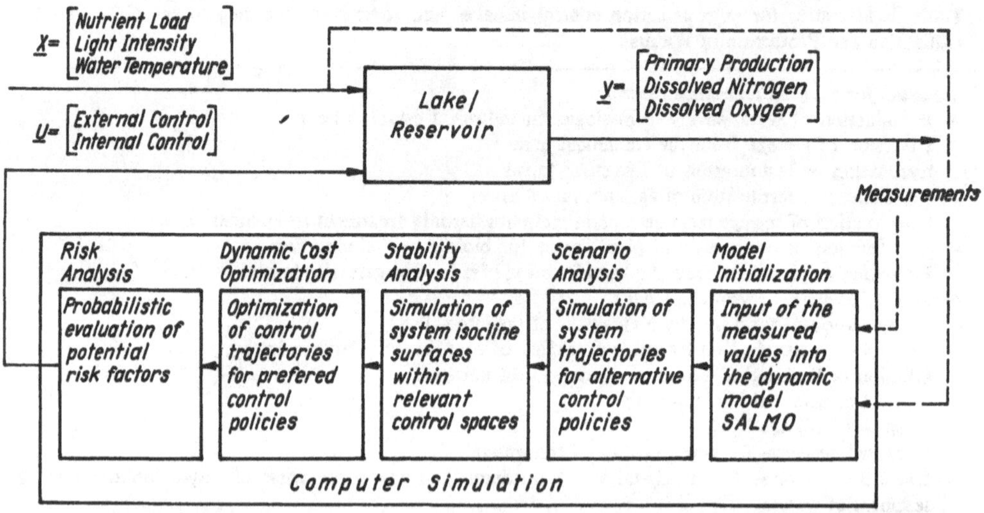

Fig. 41: Off-line approach of the model-based water quality control of lakes and reservoirs

In Sect. 7.4 a Monte Carlo simulation by means of the model SALMO is applied to assess the risk of violating the limit value of phytoplankton biomass for oligotrophic conditions. The optimal control trajectory for phosphorus elimination of a reservoir is underlied where stochastic inputs are considered.

6.1. Scenario analysis

In the meantime, scenario analysis is used by many authors as a synonym for simulation methods. Therefore, it is necessary to characterize briefly the basic assumptions of the methodics, as they are represented by Clapham, Pestel and Arnaszus (1979), Mesarovic (1979), Poole (1979) and Recknagel (1984a).

6.1.1. Method

The scenario analysis is based on a special-purpose application of the heuristic simulation method (see Sect. 2.3) to make predictions of the type „what if" using predictive dynamic models. That is, by means of scenario analysis the problem is solved: „what" kind of system behaviour appears for a definite system composition and structure and given initial condition of the state variables, „if" specified values are assigned to the input variables and parameters in the simulation time *T*. In the sense of the heuristic simulation method a gradual restriction of the control space is obtained contributing to objectify decision making.

In Fig. 42 the main steps of a scenario analysis are illustrated. Proceeding from research oriented or practice oriented problems for a system considered, at first alternative scenarios are defined based on either theoretical hypotheses or on feasible control policies. This step requires that the underlying model:

1. realistically reflects the essential state variables of the system and its interactions, without at the same time impairing its handiness and transparency
2. maintains the essential control and disturbance quantities as discrete input variables and well defined parameters.

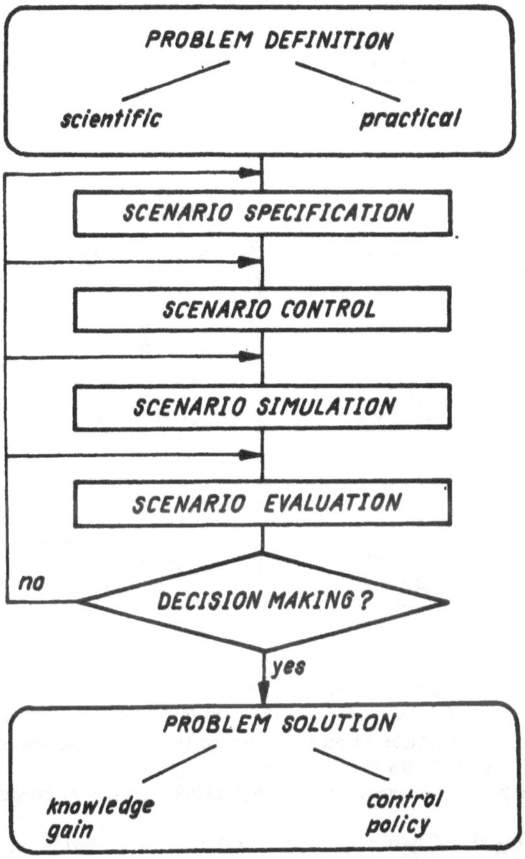

Fig. 42: General procedure of a scenario analysis

The measurable input variables and parameters of the model are potential control quantities for a scenario analysis (see also IVACHNENKO, 1982). Evaluation of the alternative scenarios is performed on the basis of relative comparison between simulated scenario trajectories and corresponding reference trajectories. This requires that the predictive validity of the underlying model is sufficiently secured. On that condition inaccuracies in the absolute values of the simulated trajectories can be tolerated (see also FEDRA, 1980). At the same time this requires the observance of the sequence of model generations according to Section 2.2, Fig. 7. A calibrated model, the predictive validity of which is not proved, can entail doubtful statements in the course of a scenario analysis.

In dependence of scenario evaluation, either the solution of the problem will be found as a scientific knowledge or a rational control policy, respectively, or scenario analysis will be repeated by modified steps.

This general run of scenario analysis stood the test when the model SALMO was used to solve scientific problems (KOSCHEL et al., 1983) and to clarify management problems (BENNDORF and RECKNAGEL, 1982; RECKNAGEL et al., 1986).

For definite waters scenario analysis begins with registration of the problems under

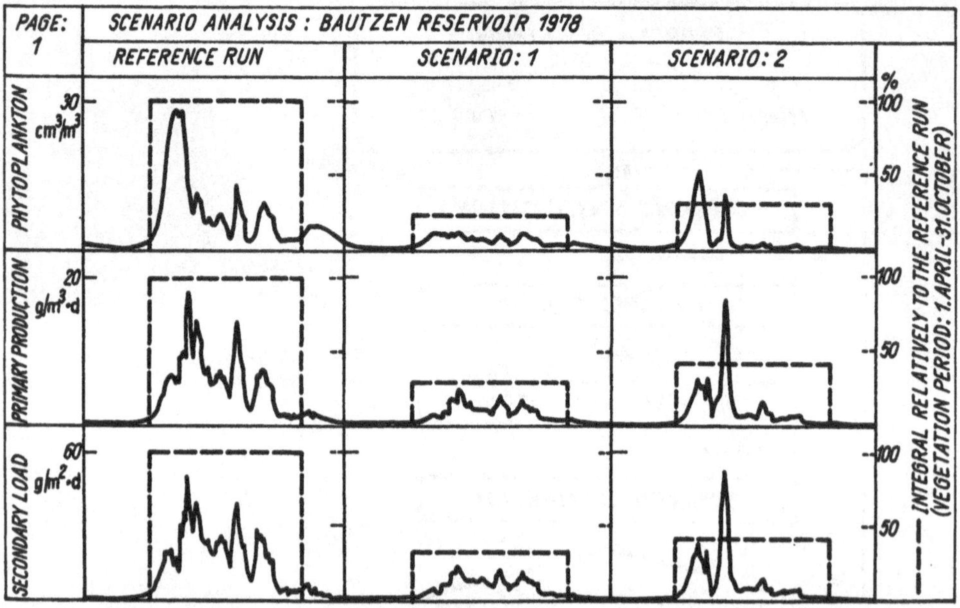

Fig. 43: Scenario analysis for the Bautzen Reservoir in 1978 (see also RECKNAGEL and BENNDORF (1982))
Scenario 1: Treatment of water inflow for reducing external phosphorus load up to 10%
Scenario 2: Manipulation of food web by predatory fish for reducing mortality of herbivorous zooplankton up to 1%. Scenario evaluation by means of comparison of trajectories and relative integrals within the vegetation period with the reference run
Page 1: Trajectories and integrals of system quantities phytoplankton, primary production and secondary load
Page 2: Trajectories and integrals of system quantities coefficient of extinction, zooplankton and mortality of zooplankton
Page 3: Trajectories and integrals of system quantities detritus and hypolimnic oxygen
Page 4: Trajectories and integrals of system quantities dissolved orthophosphate, inorganic nitrogen and N:P-ratio

consideration, scenario specification and scenario control using the model by means of so-called decision tables (e.g. see Tab. 4). In the following, on the basis of such tables the specified scenarios are simulated. The scenarios are evaluated by simulated trajectories of relevant system quantities. The trajectories are plotted in an uniform format and are supplemented by histograms (see e.g. Fig. 43). For the system quantities considered the histograms represent the surface integrals of the trajectories within the vegetation period (1st April—31th October) proportional to the reference run of 100%. The surface integrals are determined according to the Simpson rule.

The standard format of scenario analysis representation applying the model SALMO intends to consider the trajectories of the following system quantities:

— Phytoplankton total biomass
— Primary production
— Secondary load
— Coefficient of light extinction

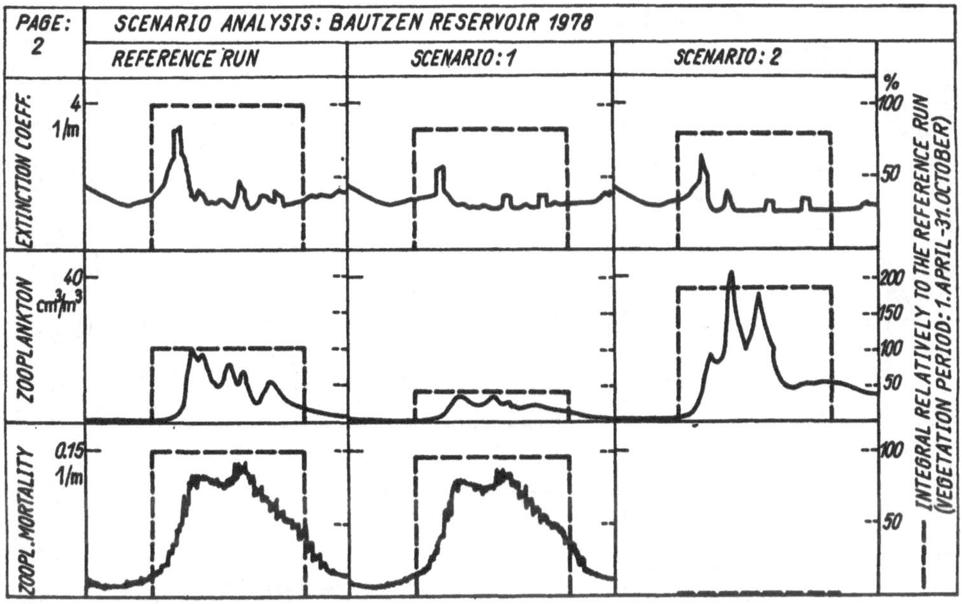

Fig. 43, Page 2

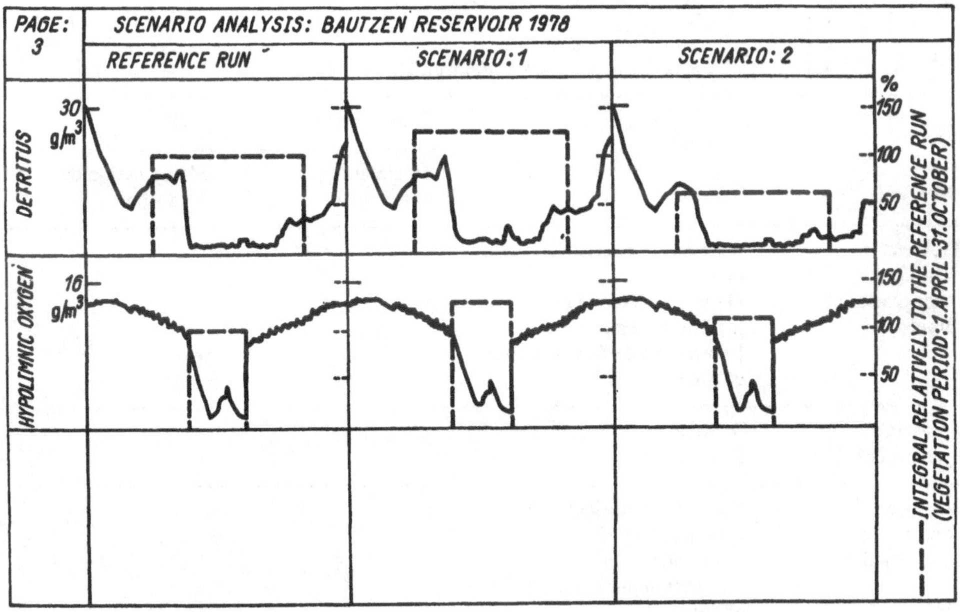

Fig. 43, Page 3

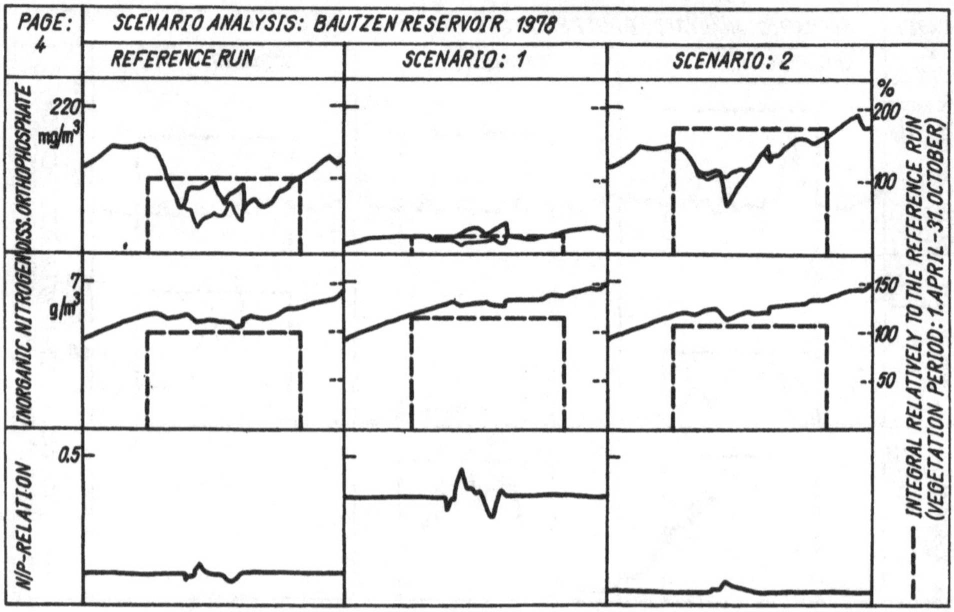

Fig. 43, Page 4

Table 4: Decision table of scenario analysis for the Bautzen Reservoir

		Control measures	
		Treatment of water inflow	Manipulation of food-web
Problem definition: management targets	Decrease in external phosphorus load	×	
	Decrease in secondary load by eutrophication	×	×
	Improvement of oxygen balance	×	×
Scenario specification	Scenario 1	×	
	Scenario 2		×
Scenario control by the model	Decrease in nutrients concentration of water inflow	×	
	Decrease in mortality of zooplankton		×

- Zooplankton biomass
- Mortality of zooplankton
- Dissolved oxygen
- Dissolved oxygen in the hypolimnion
- Dissolved orthophosphate
- Dissolved inorganic nitrogen
- N:P ratio.

During the period of stratification of the water in summer for the hypolimnion only dissolved oxygen is considered, where the integration period of the histograms refers to the specified duration of stable thermic stratification of the water and year considered. The other system quantities refer to epilimnion during the period of stratification in summer.

These system quantities extensively harmonize with the indices of water quality to classify the trophic level of standing water in accordance with the GDR-Standard on Utilization and Protection of Waters (TGL 27885/1).

6.1.2. Applications

Concerning the hypereutrophic Bautzen Reservoir by means of scenario analysis the problem was to be solved to find out whether it would be promising in eutrophication control to apply exclusively measures which reduce the external phosphorous load or whether it would be useful to take into consideration a manipulation of the food-web by predatory fish according to Shapiro, Lamarra and Lynch (1975) to reduce the mortality of zooplankton by zooplankton-feeding fish. In Tab. 4 the decision table for the scenario analysis of the Bautzen Reservoir is represented. By scenario 1 a decrease in the external phosphorous load to 10% was assumed compared against the conditions in 1978. In scenario 2 a decrease of mortality of zooplankton to 1% was defined which is obtained by an increase of predatory fish by the factor 10 and by reduction of the zooplankton-feeding fish stock by the factor 0.1 as against the conditions of 1978.

In page 1 of Fig. 43 it is shown, that in both scenarios the phytoplankton biomass is distinctly reduced (to 25—30%) and consequently, the secondary load is reduced to 40%. In page 2 of Fig. 43 it is evident that both scenarios because of the lower secondary load result in a decrease of the coefficient of light extinction to about 80% and consequently, contribute to improve the depth of transparency of the water. In scenario 1, zooplankton is reduced to 50%, whereas according to scenario 2 zooplankton biomass is nearly doubled by biomanipulation.

In page 3 of Fig. 43, for scenario 1 an increase of concentration of dissolved oxygen to 130% is simulated due to reduced zooplankton biomass and resulting zooplankton grazing. The increase in zooplankton in scenario 2 results in a reduction of dissolved oxygen concentration to 60%. By both scenarios an increase of hypolimnic oxygen to about 130% is achieved. Whereas, as expected, by scenario 1 a considerable reduction in orthophosphate concentration was achieved, in scenario 2 an increase in orthophosphate concentration to 180% is obtained by the reduced elimination performance of phytoplankton (see Fig. 43, page 4). Whereas there is no change in nitrogen concentration in the two scenarios, by scenario 1 an increase of the N:P ratio to the fourfold is obtained.

From scenario analysis for the Bautzen Reservoir it is concluded that biomanipulation

Table 5: Decision table of scenario analysis for the Bleiloch Reservoir

		Control measures		
		Advanced waste water treatment	Artificial destratification	Manipulation of food-web
Problem definition: management target	Improvement of oxygen balance	×	×	×
	Improvement of underwater light climate	×		
	Decrease in secondary load by eutrophication		×	×
Scenario specification	Scenario 1	×		
	Scenario 2	×	×	
	Scenario 3	×		×
	Scenario 4	×	×	×
Scenario control by the model	Diminution of the coefficient of extinction	×		
	Increase of mixing depth in summer		×	
	Decrease in water temperature in summer		×	
	Decrease in mortality of zooplankton			×

can be an effective measure to internal control of eutrophication of standing waters of high external phosphorus load. Since 1980, the result of scenario analysis has been used to decide principally on measures regulating fish stock in the Bautzen Reservoir (see SCHULTZ, 1983).

By scenario analysis for the Bleiloch Reservoir we aimed at finding rational management alternatives considering the different types of loads of the water. In spite of the fact that the industries in the drainage area (especially pulp processing industry) operated modern biological waste treatment plants, the following types of loads occurred in the water:

- Extremely high load of phosphorus (16.5 g/m^2 year)
- Brown colouration of the water by compounds of pulp industry which were difficult to decompose
- Disturbance of oxygen balance by a high residual organic load.

The basis for scenario analysis is an advanced treatment of waste waters planned in the catchment area resulting in an improvement of oxygen balance and — so to speak as a secondary effect — in improving under water light climate in the water. Due to the light climate improved and because of the high load of phosphorous it should be examined, whether a secondary load will be released by eutrophication which, in addition, was to be counteracted by appropriate measures (artificial destratification of water body during summer, manipulation food-web by increasing predatory fish stock).

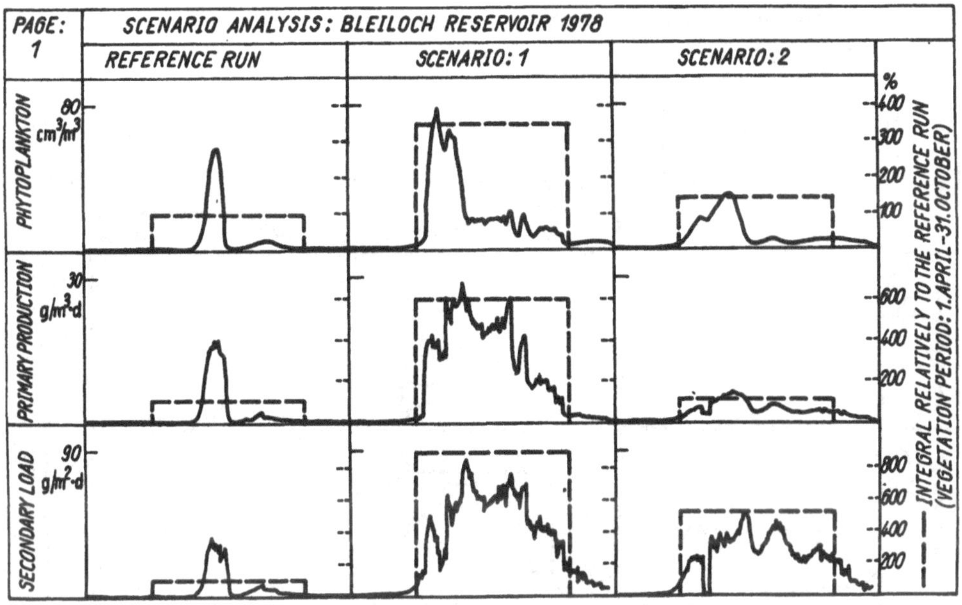

Fig. 44: Scenario analysis for the Bleiloch Reservoir in 1978 (see also Benndorf and Recknagel (1982))
Scenario 1: Advanced waste water treatment
Scenario 2: Advanced waste water treatment and artificial destratification
Scenario 3: Advanced waste water treatment and manipulation of food web
Scenario 4: Advanced waste water treatment, artificial destratification and manipulation of food web
Scenario evaluation by means of comparison of trajectories and relative integrals within the vegetation period with the reference run.
Page 1 & 4: Trajectories and integrals of system quantities phytoplankton, primary production and secondary load'
Page 2 & 5: Trajectories and integrals of system quantities coefficient of extinction, zooplankton and mortality of zooplankton
Page 3 & 6: Trajectories and integrals of system quantities hypolimnic oxygen and dissolved orthophosphate

Table 5 includes the decision table for scenario analysis for the Bleiloch Reservoir. In scenario 1 exclusively, the advanced waste water treatment was taken into consideration, where a decrease in the coefficient of extinction to 30% was assumed as against the conditions of 1978. In scenario 2, for the period of summer stagnation a combination of advanced waste water treatment and of artificial destratification was planned. By scenario 3, the combination of the advanced waste water treatment and biomanipulation to reduce mortality of zooplankton to 1% was considered as against the conditions of 1978. Finally, scenario 4 was defined to combine all three measures.

By means of page 1—3 of Fig. 44, scenarios 1 and 2 were evaluated, whereas by page 4—6 of Fig. 44 scenario 3 and 4 were evaluated. The results shown in page 1 of Fig. 44 point out that the development of phytoplankton is increased to 350% by an advanced waste water treatment further planned in scenario 1. It should be taken into consideration that in the reference run for the phytoplankton because of the high coefficient of

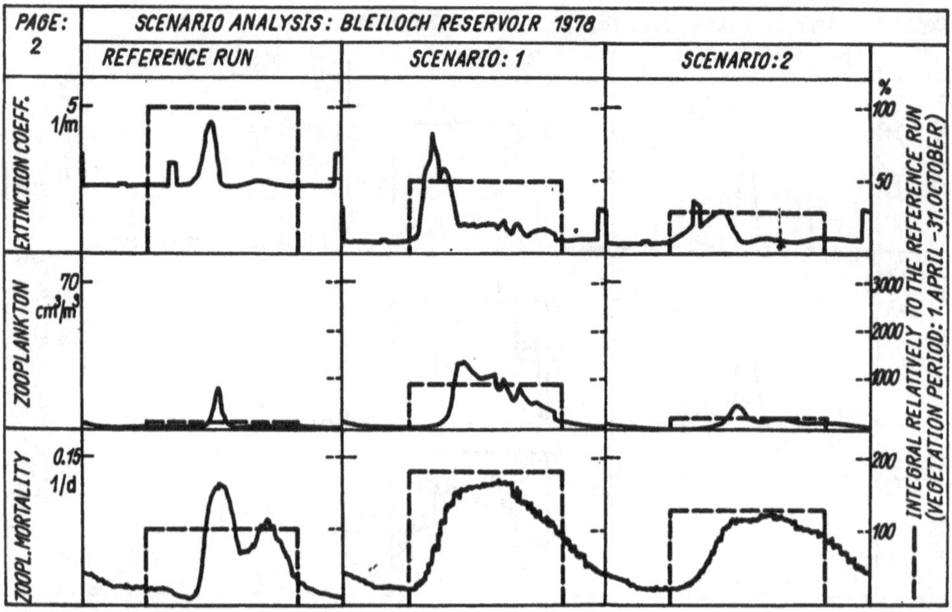

Fig. 44, Page 2

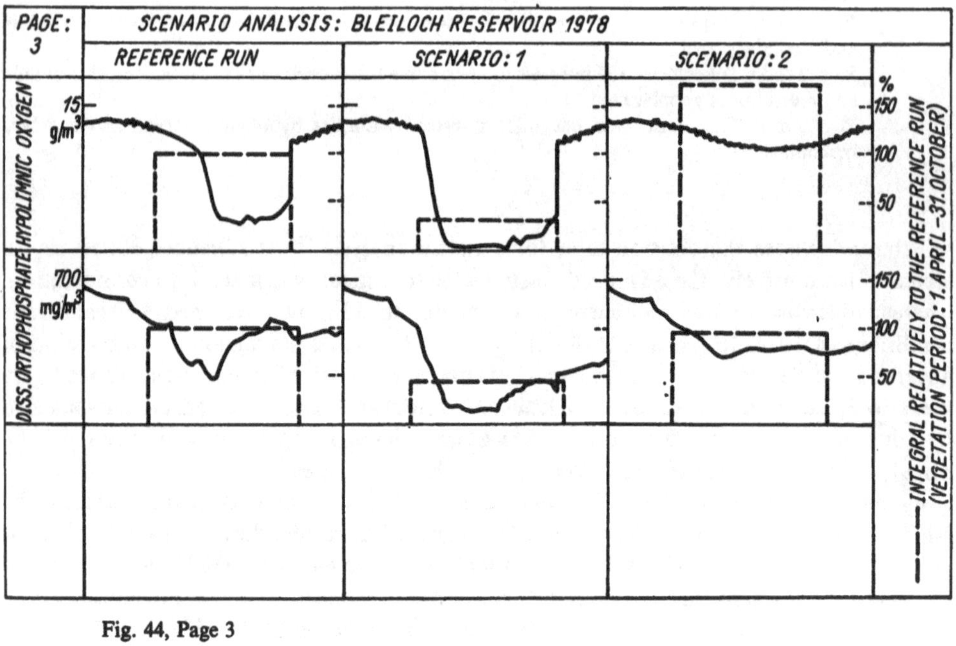

Fig. 44, Page 3

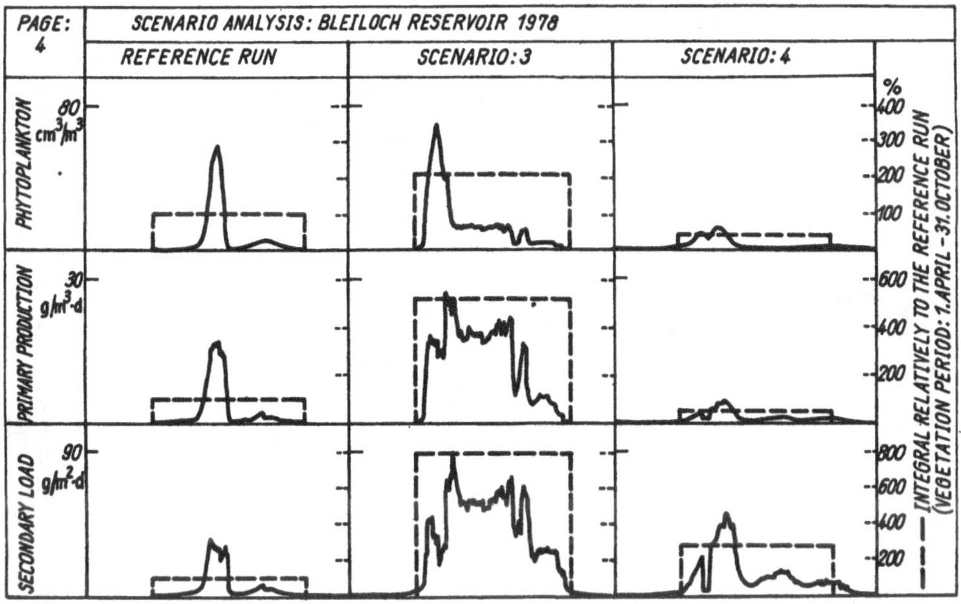

Fig. 44, Page 4

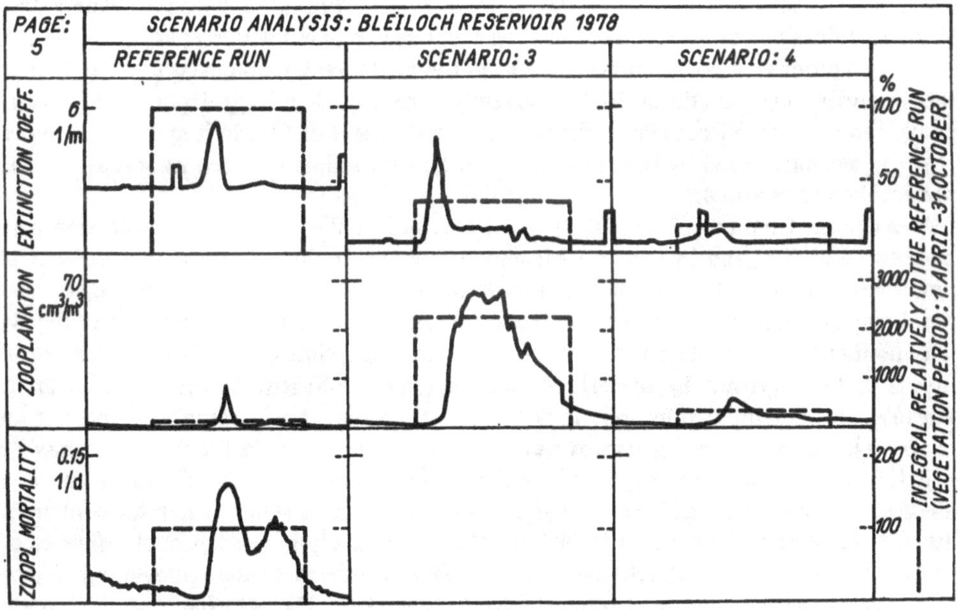

Fig. 44, Page 5

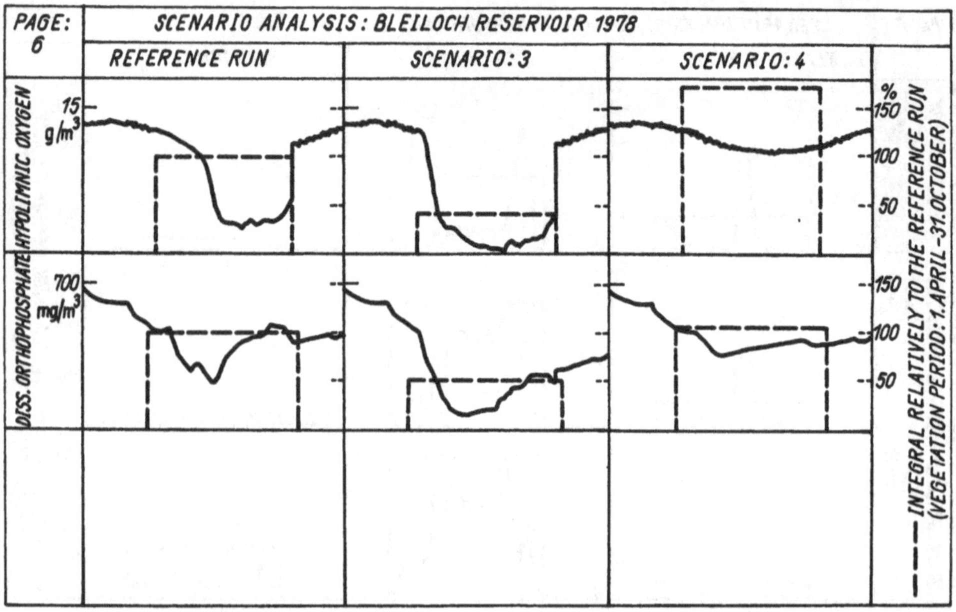

Fig. 44, Page 6

extinction and due to the brown colouration of the water only a short strong development is reduced at the beginning of summer stagnation. The increase of secondary load to 900% related with scenario 1 would bring the waters into a hypereutrophic state. However, scenario 2 would result in a distinct damping of the phytoplankton development, but would allow a secondary load of nearly 500%. In page 4 of Fig. 44 it becomes evident that scenario 3 does not cause a marked improvement as compared with scenario 1. However, scenario 4 promises a considerable decrease in the phytoplankton biomass to 50%, though the secondary load is still at 300% because of the greater mixing depth of the waters compared with the reference run. From page 1 and page 4 of Fig. 44 it may be concluded that the secondary load by entrophication can be successfully opposed by scenario 2 and preferably by scenario 4.

It is evident from page 2 of Fig. 44 that by scenario 1 the coefficient of extinction may be reduced to 50% and by scenario 2 it may be reduced to 30%, thus, by the two control policies the depth of transparency is improved. In scenario 1, because of the algal mass development, zooplankton is considerably strengthened to nearly 900%. In page 5 of Fig. 44 it is pointed out that the satisfactory under water light climate obtained by scenario 2 may still be improved by scenario 4, reducing the coefficient of extinction to 20%. Because of biomanipulation, especially in scenario 3 but also in scenario 4 there is an increase in zooplankton. To improve under water light climate in the waters, it may be concluded from page 2 and page 5 of the Fig. 44 that scenario 2 and preferably scenario 4 are used to achieve this goal. Because of missing input data on simulation of the content of dissolved oxygen in the waters for the time of scenario analysis, the aspect of influence of oxygen balance by organic residual load can only be related to the sedimenting phytoplankton biomass. It becomes obvious from page 3 of Fig. 44 that the algal mass development caused in scenario 1 results in a decrease of oxygen concentration to 30% in the hypolimnion. It can be concluded that concerning oxygen conditions the decrease in external organic load to be achieved by a further waste water treatment is again compensated by the secondary load of eutrophication. Artificial destratification of the water body during the summer period considered in scenario 2, at any rate results in a stabilization of the oxygen balance in the waters. The dissolved orthophosphate in scenario 1 is reduced to 50% by an increased elimination performance of the phytoplankton. In page 6 of Fig. 44 it becomes evident for scenario 4 that artificial destratification of the waters causes a stabilization of the oxygen balance. Whereas by scenario 3 in the period of vegetation a decrease in the orthophosphate concentration is obtained by 50% in scenario 4 the level of orthophosphate concentration remains unchanged compared with the reference run.

As to the improvement of oxygen conditions in the hypolimnion, it is recommended to realize artificial destratification of the water body planned in scenarios 2 and 4 during the summer period.

Resulting from scenario analysis for the Bleiloch Reservoir it can be stated that the measures of scenarios 2 and 4 are alternative control policies for waters, where the policy of scenario 4 should be given preference (see the criterion of secondary load in page 4 of Fig. 44). Advanced waste water treatment is an absolutely necessary measure in the sense of external preventive control. By means of artificial destratification, secondary load due to eutrophication and deterioration of the oxygen balance is efficiently counteracted.

In the sense of an internal preventive control manipulation of food-web by predatory fish additionally taken into consideration would vastly oppose to secondary load by eutrophication.

6.2. Stability analysis

Ecological systems are nonlinear dynamic systems which are extremely rich in their behaviour and tend to multiple stabilities. Multiple stabilities were observed, inter alia, for pest outbreaks in forest ecosystems (Ludwig et al., 1978), for changing species dominance in fishery systems (Jones and Walters, 1976) and in hypereutrophic ecosystems (Uhlmann, 1980). The investigation of this phenomenon is of great significance for ecosystem theory as well as ecosystem control (Fiering and Holling, 1974; May, 1977; Levin, 1979).

There are two viewpoints to investigate multiple stability: microscopic and macroscopic. The microscopic consideration can be attributed to an internal optimization problem which is inherent in each ecosystem. As Casti (1979) has stated, the ecosystem is goal-oriented and „is attempting to minimize (locally) some cost function". One of the first investigations for gaining insights into the internal optimzation of aquatic ecosystems was carried out by Radtke and Straskraba (1980).

The macroscopic view is based on stability analysis, which may be characterized by two different concepts: asymptotic and structural stability. The classical concept of the asymptotic stability was founded by Ljapunov (1949). It assumes that a system which is affected by small external perturbations will always recover to its original stable condition as soon as the perturbations are removed. So the system's behaviour would be restricted to a limited region of the state space. As pointed out by Holling (1973, 1978), Jorgensen and Mejer (1977) and Casti (1979) this concept proves to be incompatible with ecological systems. Being affected by external perturbations, the dynamic balance between maintenance and dissipation of structures generates several ecosystem steady-states which are stable (JORGENSEN and MEJER, 1977). The concept of structural stability takes into account this fact. It was introduced by ANDRONOV and PONTRJAGIN (1937) and means that a dynamic system would be said to be structurally stable if the topological properties of the phase portrait are preserved under the influence of small perturbations of the system.

The determination of boundaries and transitional regions, respectively, between the stable steady-states of an ecosystem is of considerable practical importance for the control theory. At present one way to achieve this goal will be illustrated by the application of the catastrophe theory. The basic propositions of the theory establish relationships between the dynamic structure of a system and its qualitative behaviour. The classic bifurcation theory has stressed quantitative methods for investigating structural stability, but these are generally limited to the scalar parameter case (MARSDEN and MCCRACKEN, 1976). The development of catastrophe theory to multiparameter bifurcation theory will open the quantitative approach to the multiparameter case (MEHRA, 1977; WASHBURN and MEHRA, 1980; WILSON, 1981).

This section deals with the application of catastrophe theory for a qualitative investigation of structural stability of aquatic ecosystems. Subjects of investigation are management policies for improving water quality in reservoirs which were determined in the frame of scenario analysis as outlined in Section 6.1. By this means it is aspired to obtain information for a stable control of rational management policies.

6.2.1. Concept of catastrophe theory

The catastrophe theory is a topological approach to bifurcation analysis. It was founded by THOM (1975) and further developed by ZEEMAN (1976, 1977) for investigating structural stability of dynamic systems in the state space by using the principle of topological equivalenos. For this purpose seven canonical functions were defined for so-called standard geometrical forms whose geometry and qualitative behaviour have been well studied. Suitable transformations of the system's trajectories into such a standard geometrical form make it possible to use its topological characteristics as a means to interpret the system's behaviour.

In the course of applications of the catastrophe theory three ways have been crystallized:

— Statistical matching of empirical data sets with the canonical functions
— Aggregation of existing models to the canonical functions
— Simulation of isocline surfaces by means of specific dynamic models.

The first way assumes statistical techniques such as a regression analysis which are qualified for transforming the data sets into canonical functions. Such techniques are not available at present but some efforts have been made to develop them (KEMPF, 1980; CASTI, 1981).

The second way is based on the reduction of well known laws and equations, respectively, of processes or systems under consideration to forms of the canonical functions. This way seems to be promising for ‚hard' systems such as physical ones for which deductive modelling is usual (see POSTON and STEWART, 1978; ZEEMAN, 1981; SMITH et al., 1981).

The third way at present proves to be useful for ‚softer' systems such as ecological ones for which inductive modelling is preferred. On the basis of dynamic models which have been

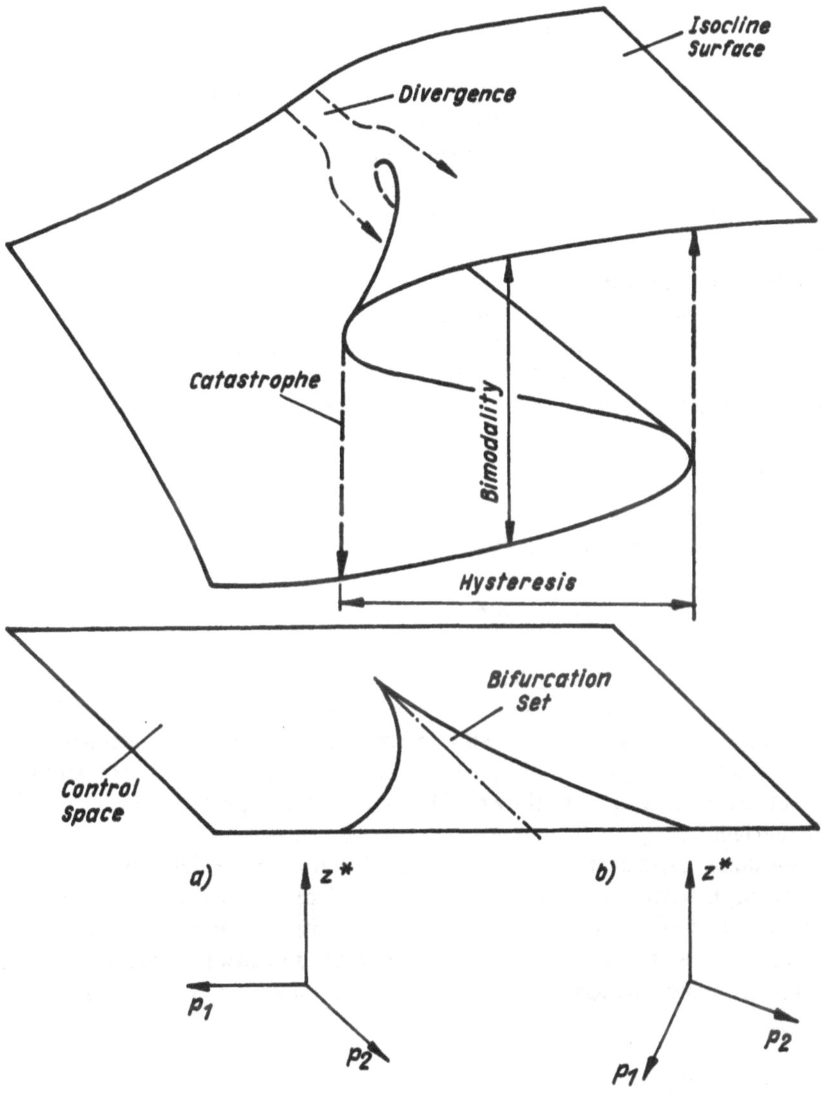

Fig. 45: Standard geometrical form of the cusp catastrophe. p_1, p_2, control parameters; z^*, set of equilibrium points,
a) p_1 acts as normal factor, p_2 acts as splitting factor
b) p_1 and p_2 act as conflicting factors

validated to the necessary extent simulation of the systems's trajectory family in the state space is possible. Structural stability can be investigated by an optical comparison with the standard geometrical forms (see JONES, 1977; RENGUET and DUBOIS, 1981).

The investigation of structural stability of ecosystems primarily serves to clarify the consequences of anthropogenous perturbations (e.g. control measures) or natural perturbations (e.g. meteorology) on ecosystem behaviour. For that reason it is sensible to embed the catastrophe theory into scenario analysis based on a simulation model. This aspect favours the third way of applying the catastrophe theory to aquatic ecosystems pursued in the next section.

The essential principles of catastrophe theory will be discussed briefly. The most relevant geometrical form at hand, namely the cusp is shown in Fig. 45. It is named after the cusp-shaped bifurcation set which is obtained from the projection of the isocline surface into the control space. The isolation surface is determined by the set z^* of the equilibrium points z^* of the so-called behaviour variables z which are generally considered to be state variables of the system.

Referring to equation $dz/dt = f(z, p)$, z^* is obtained for fixed p by $f(z^*, p) = 0$, where p means the set of control parameters.

In the case of the cusp catastrophe only two control parameters, p_1 and p_2, are considered which decisively influence the positions of the points z^*.

The canonical function F for the cusp reads as follows:

$$F(p_1, p_2, z) = \frac{z^4}{4} - p_1 z - \frac{p_2 z^2}{2}.$$

Thus, the isocline surface is given by:

$$\frac{\partial F}{\partial z} = 0, \qquad z^3 - p_1 - p_2 z = 0.$$

Depending on their nature, the control parameters can have different topological positions in relation to the cusp-axis. As represented in Fig. 45a, the control parameter p_1 stands vertically to the cusp-axis while p_2 is directed along the cusp-axis. In this case, p_1 is called the *normal factor* while p_2 is the *splitting factor*. If the control parameters lie on either side of the cusp-axis, as shown in Fig. 45b, the two parameters are considered to be conflicting factors.

The idealized cusp catastrophe is characterized by four properties: bimodality, catastrophe (discontinuity), hysteresis (delayed response) and divergence (see Fig. 45). A common interpretation of this is given by ZEEMAN (1977); its relevance for ecological systems is discussed by JONES (1977). A survey of higher dimensioned catastrophes and their applications is given by ZEEMAN (1977) and POSTON and STEWART (1978).

6.2.2. Method

The third way of application of the catastrophe theory was used to investigate structural stability, as discussed in the previous section. With the help of the model SALMO isocline surfaces of significant biological state variables were simulated in the relevant control space. Each isocline surface was generated by equilibrium points of the system. The equilibrium points were calculated on the basis of mean biomass integrals of the annual vegetation period. The control space was defined by the 19×9-matrix to represent relevant combinations of two control parameters. To make the transitional regions between different attractor (or structurally stable) regions of the isocline surface more transparent the projections of ad hoc defined isolines into the control space are depicted. The isolines refer to constant biomass gradients given in cm^3/m^3. Changes of the control parameters are considered relatively to their nominal level (100%).

6.2.3. Applications

The scenario analysis for the hyper-eutrophic Bautzen Reservoir has shown that an efficient measure for eutrophication control of hypereutrophic lakes and reservoirs can be offered by biomanipulation (see Section 6.1). In the context of this section biomanipulation means artificial increase of predatory fish stock in order to decrease zooplankton-eating fish stock and consequently to decrease zooplankton mortality. Phytoplankton grazing by herbivorous zooplankton thus becomes an efficient internal mechanism for eutrophication control. In terms of the control theory, this measure requires a feedback control because it is dependent on the actual state of the system. Therefore it is necessary to analyze the

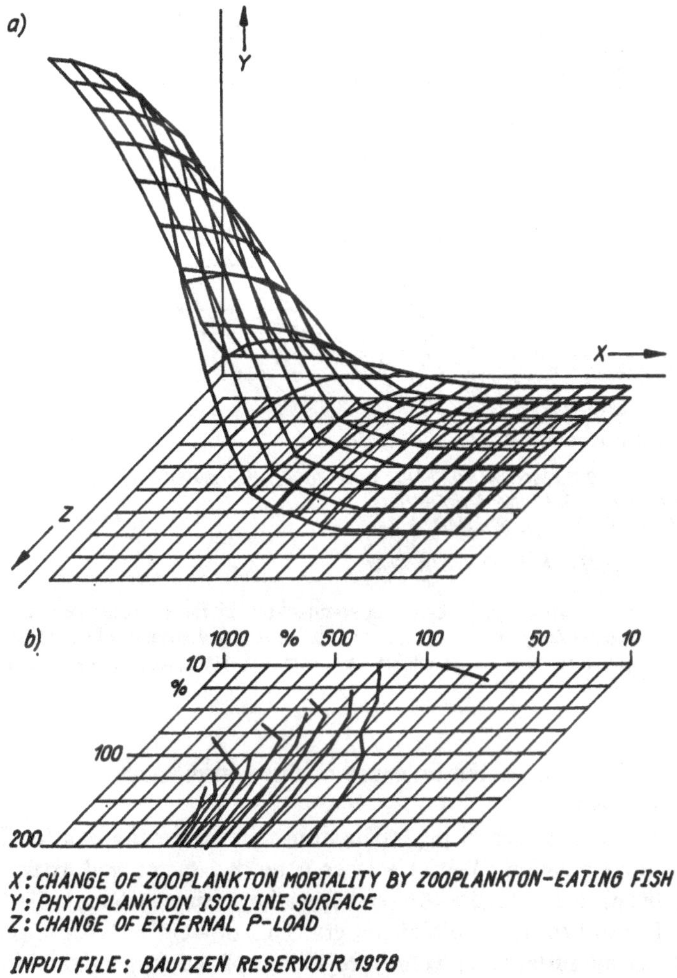

Fig. 46: Structural stability of phytoplankton (diatoms, Chlorococcales) in the relevant control space of the Bautzen Reservoir, a) Phytoplankton isocline surface, b) Isoline projections into the control space. Isolines correspond to biomass gradients of 4 cm^3/m^3; 100% corresponds to the nominal control level in 1978

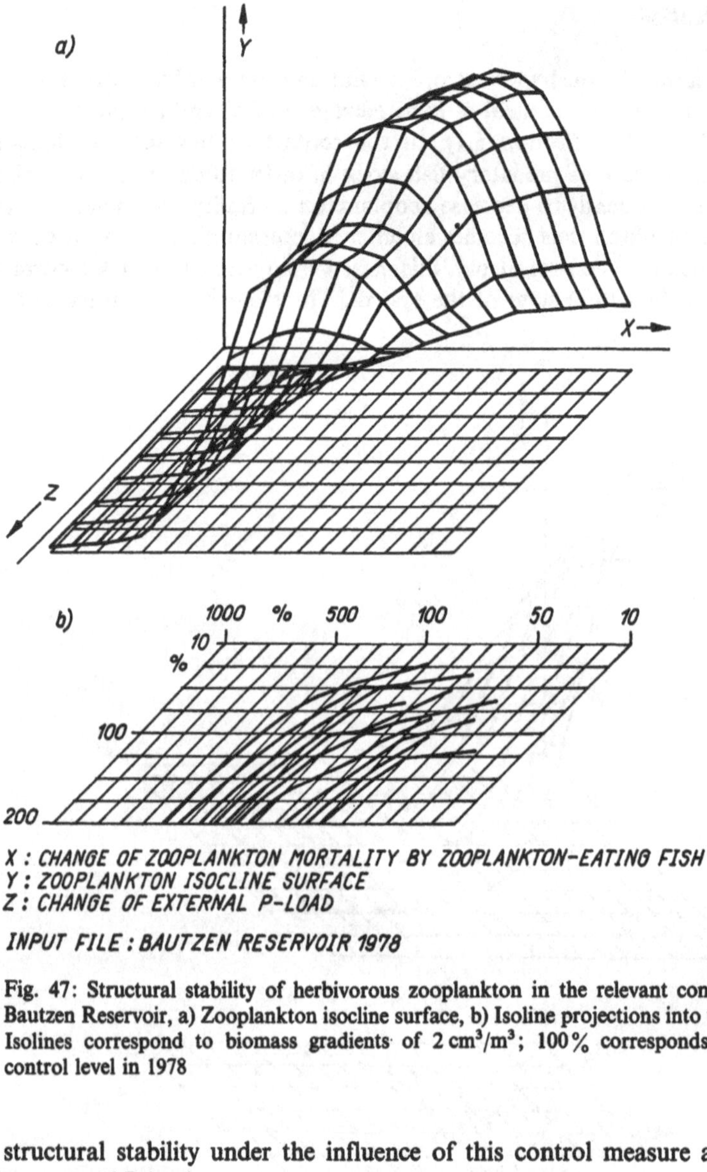

Fig. 47: Structural stability of herbivorous zooplankton in the relevant control space of the Bautzen Reservoir, a) Zooplankton isocline surface, b) Isoline projections into the control space. Isolines correspond to biomass gradients of 2 cm^3/m^3; 100% corresponds to the nominal control level in 1978

system's structural stability under the influence of this control measure and the effects of changing external P-load.

In Fig. 46a the isolation surface of phytoplankton biomass for the Bautzen Reservoir is shown. Here, biomanipulation (X-axis) acts as a normal factor and external P-load (Z-axis) acts as a splitting factor. Figure 46b represents the pertinent isoline projections into the control space. The isolines refer to biomass gradients of 4 cm^3/m^3. Corresponding to Fig. 46, in Fig. 47 isocline surface and isoline projections of the zooplankton biomass are depicted. Zooplankton biomass gradients of the isolines amount to 2 cm^3/m^3.

The analysis of Figs. 46 and 47 indicates that the classical cusp catastrophe in system behaviour characterized by the cusp-shaped bifurcation set does not occur in the control space. But it is evident that there is a tendency to very fast changes between strongly divergent, bimodal regions of system behaviour which are structurally stable. To meet the management goal of ‚eutrophication control' for the Bautzen reservoir by the discussed measures care should be taken that zooplankton mortality due to zooplankton-eating fish is not higher than 70% of the nominal level in 1978 (100%). It will then be possible to compensate phytoplankton growth in a stable manner by zooplankton grazing up to a doubling of the external P-load (see Fig. 46). Field investigations on eutrophication control of the Bautzen Reservoir by biomanipulation performed since 1978 are confirming these general conclusions but moreover new behavioural effects have been perceived, which at present are beyond the model structure of SALMO (see BENNDORF et al., 1983).

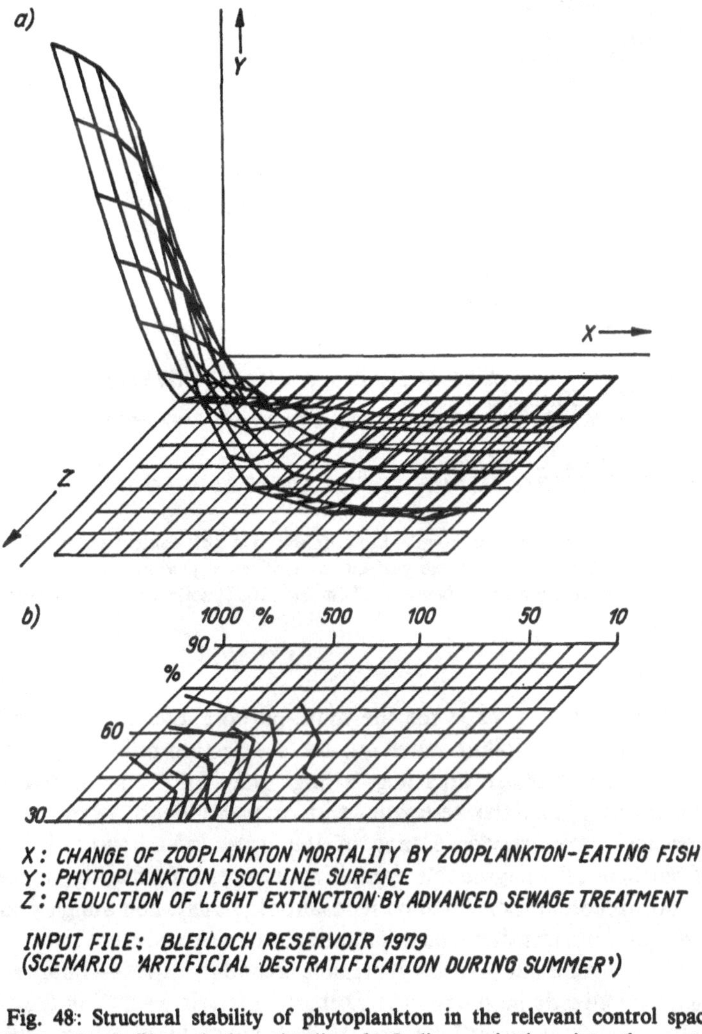

Fig. 48: Structural stability of phytoplankton in the relevant control space of the Bleiloch Reservoir. a) Phytoplankton isocline, b) Isoline projections into the control space. Isolines correspond to biomass gradients of 2 cm^3/m^3; 100% corresponds to the nominal control level in 1978

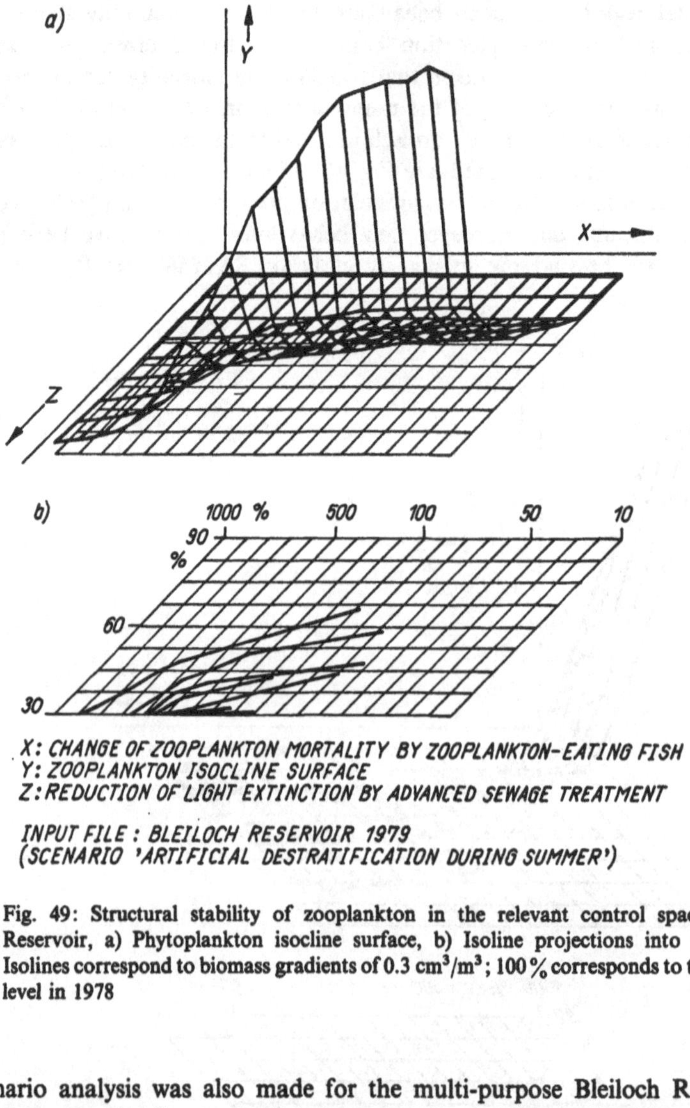

Fig. 49: Structural stability of zooplankton in the relevant control space of the Bleiloch Reservoir, a) Phytoplankton isocline surface, b) Isoline projections into the control space. Isolines correspond to biomass gradients of 0.3 cm^3/m^3; 100% corresponds to the nominal control level in 1978

A scenario analysis was also made for the multi-purpose Bleiloch Reservoir to find management strategies for controlling water quality (see Section 6.1). In spite of the fact that the industries in the drainage area (e.g. a large pulp mill) operate two-stage biological waste treatment plants the reservoir, with a mean depth of 40 m, length of 20 km and theoretical mean residence time of the water of 0.9 year, is subject to: residual load of biochemically degradable organic substances affecting the oxygen budget of hypolimnion; brown colouration (mainly lignosulfonic acids); and a high phosphorous and nitrogen load, also from transboundary domestic sources. An advanced waste treatment for improving oxygen balance and — as an inevitable secondary effect — improving underwater light climate had already been designed. Scenario analysis resulted in the conclusion that, owing to improved underwater light climate and due to high orthophosphate load, eutrophication is enhanced to an extent sufficient to compensate for the effects of the measures mentioned above. Thus a rational management policy should be a combination of artificial destratification during summer and biomanipulation in addition to the advanced waste treatment.

In Figs. 48 and 49, the isocline surfaces of phytoplankton and zooplankton biomass are shown for the Bleiloch Reservoir as well as isoline projections into the control space. Biomass gradients were defined as 2 cm^3/m^3 and 0.3 cm^3/m^3, respectively. As control parameters the change of zooplankton mortality caused by zooplankton eating fish (X-axis) and reduction of light extinction by advanced waste treatment (Z-axis) have been selected. Artificial destratification of the water body during the summer months is permanently simulated by a modified input file.

The isocline surface in Figs. 48 and 49 show no cusp catastrophe in the relevant parameter space considered but only one stable region as well as one region with very fast changes. The region with strong changes in the phytoplankton isocline surface (Fig. 48a) is determined by improving underwater light climate by advanced waste treatment. Reduction of the light extinction below 70% involves the danger of substantial phytoplankton outbreaks by eutrophication, but this should be controllable by biomanipulation of zooplankton mortality caused by zooplankton-eating fish not higher than 90% of the nominal level Figs. 48b and 49).

6.2.4. Conclusions

The simulation of isocline surfaces as one way of applying the catastrophe theoretical concept has proved to be suitable for investigating structural stability of aquatic ecosystems. The approach was applied as an aid in policy making for water quality management of reservoirs in the German Democratic Republic. On the basis of rational management policies derived from scenario analysis, relevant information may be obtained which is necessary for a stable control. For this purpose an attempt was made to quantify the qualitative behaviour of aquatic ecosystems relatively to their nominal control levels in terms of control parameters.

6.3. Optimal control

In Section 2.4 a general characterization of the control theory was pointed out. The solution of the objective function defined in (1) of an optimal control always implies an optimization problem. It is solved either by a scalar optimization or by a multicriteria optimization depending on whether the special control problem includes one or several objective functions. Control problems including several objective functions often are solved approximately by scalar optimization in maintaining only one objective function and the remaining ones are taken into consideration as state space constraints.

To control water quality in lakes and reservoirs, at least two objective functions are required (see chapter 1):

— a maximization of water quality
— a minimization of costs.

In the following, ways are pointed out to solve this control problem by means of *scalar optimization with state space constraints* and by *multicriteria optimization*. The optimization technique in general is based upon a combination of optimization and simulation methods. Depending on whether the human being is actively included or not, a distinction is made between interactive or completely algorithmic optimization techniques. The scalar optimization is completely algorithmically solved, whereas the multicriteria optimization is presented by an interactive technique.

6.3.1. Scalar optimization with state space constraints

According to GOH (1979) the scalar optimization for a continuous control of a dynamic system with a finite control time T is represented as follows:

dynamic system:	$\dot{z}_i = f_i(\boldsymbol{Z}, \boldsymbol{U})$,	$i = 1(1)\,n$
initial condition:	$z_i(0) = z_{i0}$	
final condition:	$\Phi^s(T, \boldsymbol{Z}(T)) = 0$,	$s = 1(1)\,m$
control constraints:	$a_j \leqq u_j \leqq b_j$,	$j = 1(1)\,k$
state space constraints:	$f_h(\boldsymbol{Z}(t), \boldsymbol{U}) \leqq 0$,	$h \leqq \{1, \dots, n\}$
	$0 \leqq t \leqq T$,	

objective function:

$$\int_0^T I(\boldsymbol{Z}, \boldsymbol{U}, t)\, \mathrm{d}t \rightarrow \min_{\boldsymbol{U}}$$

where:

$\dot{z}_i$: trajectory of the ith state variable
z_{i0}: initial value of the ith state variable
Φ^s: mapping operator
a_j: lower constraint of the jth control variable
b_j: upper constraint of the jth control variable
I: index of optimality.

The index of optimality I is represented by a scalar function.

To apply this formalism to water quality control, at first an appropriate objective function should be defined where further objectives can be taken into consideration as state space constraints. To do this, different approaches are feasible. Optimizing the Lake Biwa (Japan) MATSUMURA and SAKAWA (1981) do not consider the costs. As an objective function, they define the minimization of phytoplankton biomass by optimally controlling the external nutrient load. The minimally occuring and maximally admissible nutrient loads are defined as control constraints, therefore, no state space constraints are required.

HUGHES (1982) defined minimization of the weighted sum of deviations from nutrients load given in advance as an objective function of an optimal control for the Lake Balaton, where the absolute limitation of investment and operation costs to a budget available as constraint is taken into consideration. To control water quality, this solution is of a high relevance, as limitation of costs is more realistic than the use of absolute water quality limits or load limits, respectively, considering dynamics inherent to the aquatic ecosystem.

In the paper of CALCEVA et al. (1982) the objective function aims at minimizing total costs where the observance of an absolute limit for phytoplankton biomass is defined as state space constraint. The solution we used to optimally control water quality by scalar optimization orients to minimize the costs function, where the state space constraint for phytoplankton biomass is seasonally modified.

6.3.1.1. Definition of the problem

Corresponding to the course of an integrated decision making according to Fig. 41, the most prevential control policy should be determined from the control alternatives which has proved to be rational. Those investment costs should not exceed a certain budget, the annually occuring operation costs are reduced to a minimum and seasonally specified water quality limits are not exceeded. The optimization model required should be flexibly designed in order to be applicable for optional waters and can be modified for significant alternative control policies. For this aim, at first the following simplifying assumptions should be made:

- The control period T applies only to one year, consequently, only such control measures should be taken into consideration as dynamic control variables whose technology or mode of action allow an operational control and the effects of which on the system behaviour are perceptible only at a minor temporal delay. For example, in order to optimally control the measure biomanipulation, in future, a control period of at least 5 years should be examined. So far, such measures in the model are controlled by constant values for the period of one year.
- For dynamic control, four control variables were defined as usually, so that they can be combined selectively to optimize an actual control policy. In the pertinent cost functions, only linear dependencies of operation costs on control variables are assumed.
- To optimize continuous control of a dynamic system, an approximation by equidistant, discrete control spans within the control period T is required. The length of the discrete control spans can be optionally selected, however, it has to correspond to a justifiable computing time.

For dynamic optimization, so far the following control variables are planned to be applied:

1. Decrease of external orthophosphate load

This control variable refers to a P-elimination plant according to the so-called Wahnbach method (Bernhardt and Clasen, 1981). By this method, external P-load is decreased by about 90% by chemical precipitation, flocculation and filtration, where the efficiency factor of the method is controlled by the filter running time which depends on energy supply. The different operation costs of the method result from the amount of precipitants and energy required to obtain a definite efficiency factor. For the Wahnbach method, the control variable u_W is defined as follows:

$$a_W \leqq u_W(t) \leqq b_W\,, \qquad 0 \leqq t \leqq 360$$

$$\mathrm{PIN}_W(t) = \mathrm{PIN}(t) \cdot (1 - u_W(t))\,.$$

Costs function:

$$c_W = \mathrm{IC}_W + \sum_{t=t+\Delta t}^{360} \frac{u_W(t) \cdot \mathrm{OC}_W}{b_W}$$

with $\Delta t = 1$ d where:

$\mathrm{PIN}(t)$: orthophosphate concentration in the inflow of waters in the uncontrolled case

$\mathrm{PIN}_W(t)$: orthophosphate concentration decrease in the inflow of waters according to the efficiency factor controlled by the Wahnbach method

IC_W: costs of investment for the P-elimination plant according to the Wahnbach method

OC_W: operation costs of the Wahnbach method at a maximum efficiency factor of 90% P-elimination referred to the selected discrete control span.

2. Artificial destratification

Artificial destratification is used in the period of stable thermic stratification of deep waters. As a rule, this measure is realized according to the „air-lifting" principle (KLAPPER, 1977) where at the deepest point of water compressed air escapes from corresponding distributing devices and thus cold deep water is moved into the epilimnion. With progressing process hypolimnic and epilimnic waters are completely mixed, thus, stratification is annulled. This results in an approximately homogenous distribution of temperature, nutrients and oxygen in waters. Because of the mixing depth simultaneously increased, specific light consumption of phytoplankton is decreased and consequently, secondary load of waters is reduced by eutrophication.

For artificial destratification the control variable u_D is defined as follows:

$$u_D(t) = \begin{cases} 0 \\ 1 \end{cases}, \quad STS \leqq t < VZH$$

with STS: 1st decade of summer stagnation of the water

VHZ: 1st decade of autumn complete circulation of the water.

The control of this action is only stated in the case $u_D(t) = 1$. There, the following input variables are modified:

$$\begin{aligned} V &= V_D \\ ZMIX &= ZMIX_D \\ ZMIXREAL &= ZMIXREAL_D \\ QIN &= QIN_D \\ QOUT &= QOUT_D \\ T &= T_D \end{aligned}$$

Costs function:

$$c_D = IC_D + \sum_{t=STS+\Delta t}^{VZH} OC_D\, u_D(t)$$

with $\Delta t = 10$ d where

V: water volume of total water body

ZMIX: mean mixing depth of water referred to the total water body

ZMIXREAL: actual mixing depth referred to the total water body

QIN: water inflow referred to the total water body

QOUT: water outflow referred to the total water body

T: mean water temperature referred to the total water body

IC_D: costs of investment for the technology of artificial destratification

OC_D: operation costs to control artificial destratification referred to a decade

Optimization by control variable u_D represents an all integer problem requiring appropriate modifications in the optimization algorithm (see also HUGHES, 1982).

As to control variable u_D for artificial destratification, so far the following simplifying assumptions are made:

- the measure should only be taken to prolongate the complete circulation of spring time by decade into the summer period; so far, an operative destratification is impossible after temporary formation of a stable stratification
- the measure is simulated so that without any time delay immediately destratification is the result
- operation costs are not differentiated independently of the fact whether the density gradient of stratification still exists or was already surmounted.

3. Decrease of orthophosphate concentration in the water by chemical precipitation

This measure is taken by using ferric chloride or aluminium chloride and results in orthophosphate deposit in the water's sediment by chemical bonding of orthophosphate (see KLAPPER, 1977).

For chemical precipitation the control variable is defined as follows:

$$a_P \leqq u_P(t) \leqq b_P\,, \qquad 0 \leqq t \leqq 360$$

$$\mathrm{PSED}_P(t) = \mathrm{PSED}(t) \cdot u_P(t)$$

$$\mathrm{XSED}_P(t) = \mathrm{XSED}(t) \cdot u_P(t)\,.$$

Costs function:

$$c_P = \sum_{t=t+\Delta t}^{360} \frac{u_P(t) \cdot \mathrm{OC}_P}{b_P}$$

with $\Delta t = 1$ d where:

PSED(t): sedimentation of orthophosphate in the uncontrolled case
$\mathrm{PSED}_P(t)$: sedimentation of orthophosphate increased by chemical precipitation
ISED(t): sedimentation of phytoplankton in the uncontrolled case
$\mathrm{ISED}_P(t)$: increased sedimentation of phytoplankton by coprecipitation
OC_P: operation costs of chemical precipitation at maximum admissible addition of precipitation agents.

4. Dilution of standing waters by water poor in nutrients

This measure requires to dispose of a storage system composed by several basins providing sufficient water poor in nutrients for the main basin from a distance which is justifiable (KLAPPER, 1977). This measure can be controlled to increase either the water depth in reservoirs or to flush basins of reservoirs and seas by decreasing the mean residence time of water. Applying this measure, the quantity of water with respect to the required quantity of water outlet for uses is to be considered, so in future, this measure is to be combined with optimization of the outlet of water quantities of a storage system. At present, combined water quantity/quality simulation of storage systems is prepared, first preparations were done by SCHEIN (1982) and SIX (1983).

Considering the prerequisites mentioned control of this measure is defined as follows:

$$a_V \leqq u_V(t) \leqq b_V\,, \qquad 0 \leqq t \leqq 360$$

$$\mathrm{QIN}_V(t) = \mathrm{QIN}(t) \cdot u_V(t)$$

$$\mathrm{QEIN}_V(t) = \mathrm{QEIN}(t) \cdot u_V(t)$$

$$\mathrm{QHIN}_V(t) = \mathrm{QHIN}(t) \cdot u_V(t)\,.$$

In addition, control of the basin's flushing is taken into consideration as follows:

$$\mathrm{QOUT}_V(t) = \mathrm{QOUT}(t) \cdot u_V(t)$$

$$\mathrm{QHOUT}_V(t) = \mathrm{QHOUT}(t) \cdot u_V(t)\,.$$

Costs function:

$$c_V = \sum_{t=t+\Delta t}^{360} \frac{u_V(t) \cdot \mathrm{OC}_V}{b_V}$$

with $\Delta t = 1$ d;

QIN(t):	water inflow in the total water body of water
$\mathrm{QIN}_V(t)$:	water inflow increased into the total water body by dilution water poor in nutrients
QEIN(t):	analogously to QIN(t) referred to the epilimnion of water
$\mathrm{QEIN}_V(t)$:	analogously to $\mathrm{QIN}_V(t)$ referred to the epilimnion of water
$\mathrm{QHIN}_V(t)$:	analogously to $\mathrm{QIN}_V(t)$ referred to the hypolimnion of water
QOUT(t):	water outflow from the total water body of water
$\mathrm{QOUT}_V(t)$:	increased water outflow from the total water body due to flushing of the storage basin
QHOUT(t):	water outflow from the hypolimnion of water
$\mathrm{QHOUT}_V(t)$:	increased water outflow from the hypolimnion of water due to flushing of the storage basin.

To dynamically control the water quality of standing waters the problem of scalar optimization is summarized as follows:

System: $\dot{z}_i = f_i(\mathbf{Z}, U), \qquad i = 1(1)\,4$

with z_1: orthophosphate
z_2: phytoplankton group 1
z_3: phytoplankton group 2
z_4: zooplankton

initial condition: $z_i(0) = z_{i0}$

constraint of the control: $a_j \leqq u_j(t) \leqq b_j\,,$
$j = 1(1)\,n\,, \qquad n \leqq 4\,, \qquad t = 0(1)\,360$

state space constraint: $z_2(t) + z_3(t) \leqq g(t)$

object function

$$C \to \frac{\min}{U} \quad \text{with} \quad C = \sum_{j=1}^{n} c_j(u_j)\,.$$

Here:

$g(t)$: limit of total phytoplankton biomass according to the national water quality standard

c: total costs of dynamic control of n control variables

c_j: costs of dynamic control of the jth control variable.

To solve the problem when the state space constraints are taken into consideration a modified Lagrange method according to WIERZBICKI and KURCYUSZ (1977) is applied (cf. KALCEVA et al., 1982). In this connection, an important partial problem occurs:

$$\frac{\min}{U} L_r(U, \lambda)$$

where the function L_r is defined as follows:

$$\begin{aligned} L_r(U, \lambda, t) = C - \int_0^{360} \max \left\{ z_2 + z_3 - G, \frac{\lambda}{2r} \right\} dt \\ + r \int_0^{360} \max^2 \left\{ z_2 + z_3 - G, \frac{\lambda}{2r} \right\} dt \end{aligned} \tag{6}$$

L_r: the modified Lagrange function

λ: vector of the Lagrange multipliers

r: Penalty factor.

6.3.1.2. Solution

The optimization problem was solved by combining the optimization program DNFMF 2 (SCHWETLICK et al., 1977) and the simulation model SALMO according to the flow chart in Fig. 50.

The subroutine DNFMF 2 is based upon a regularized Newton method without derivatives to minimize a function. The function used was a Lagrange function calculated according to (6). For particular iteration steps this function is based upon the state trajectories simulated by SALMO. Thus, calculation of partial derivatives of the considerably nonlinear state equations with respect to changes of control variables was avoided which would be indispensible for an optimization method with derivatives such as the method of conjugated gradients (see KALCEVA et al., 1982). Partial derivatives are always determined approximately by divided difference gradients methods, thus additional inaccuracies during optimization are induced. At present, much time spent on computing required by various iterations restricts the approach of this problem. By the ROBOTRON-computer 1022, a CPU time of 60 minutes is required to determine about 40 values of function for one iteration step of a control variable with 6 discrete time intervals within the control period T of one year. Every increase of iteration steps results in a further approximation of the optimum solution. A way out would be the availability of the parallel processor technique for simulation (see Section 2.3) which would considerably contribute to faster calculations of state trajectories of SALMO. At present the solution of the optimization

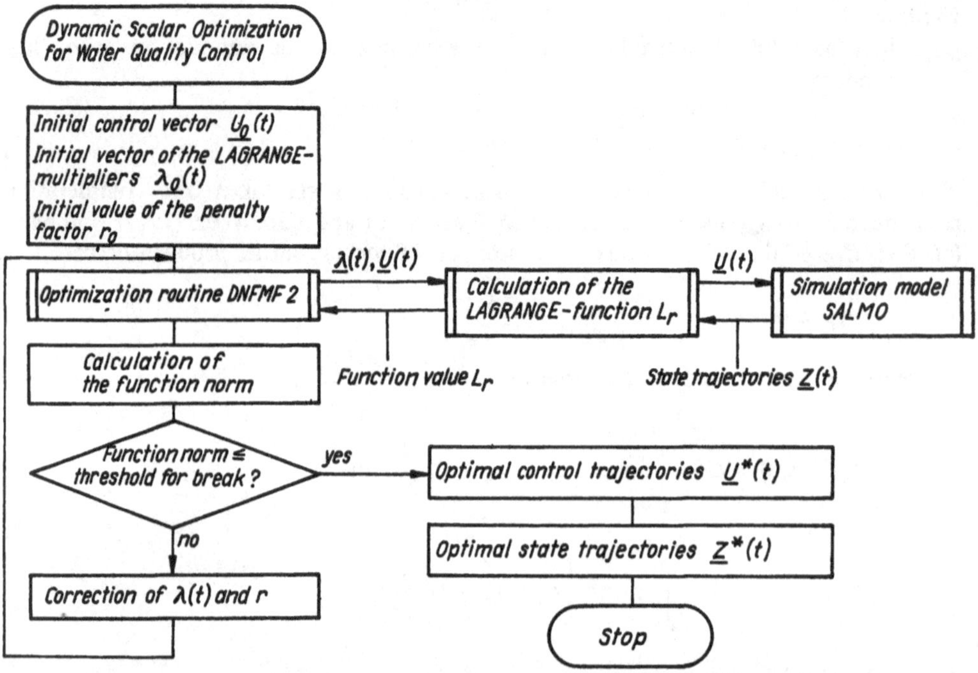

Fig. 50: Flow chart of the program for dynamic scalar optimization with state space constraints

problem by a gradient method is pursued, where a compromise should be found between accuracy of the solution and a justifiable computing time.

First results in applying the program system according to the flow chart in Fig. 50 for optimal control of water quality of the Eibenstock Reservoir are presented in Chapter 7. For numerical integration of the state equations of SALMO the Runge-Kutta method of 4th order with an interval of one day is used. The integrals on the state trajectories in (6) are determined by the Simpson rule. Control limits of the optimization problem are considered by corresponding transformation equations.

6.3.2. Multicriteria optimization

Multicriteria optimization of a control problem can be represented as follows:

$$\min_{U} f(U), \qquad U = (u_1, u_2, \ldots, u_m) \tag{7}$$

$$g_i(U) \leqq 0, \qquad i = 1(1)\,n$$

where:

f_i: the ith objective function, $f = (f_1, \ldots, f_m)$
g_i: the ith constraint function.

According to SEO and SAKAWA (1979) multicriteria optimization is characterized by the following difficulties:

1. For all control variables the best solution can not be found, as they are in conflict with each other.
2. In general, the different objective functions have no uniform magnitude, their numerical values can not be compared.
3. Inaccuracies inherent in the solutions of multicriteria problems at drawing up of decisions can not or only determined with difficulties and must be taken into consideration.

To overcome the difficulty mentioned in 1., it is aimed at finding a preferable solution within a Pareto optimal solution set. The difficulty mentioned in 2. may be overcome in setting up a common evaluation index for all objective functions, expressing a preference value of the decision maker for the particular objective function. However, it is difficult and risky to define an overall evaluation index, subjective inaccuracies may be induced or changes in the objectives may be occur in the course of the decision making.

Consequently, optimization of a multicriteria problem includes two general steps:

1. determination of the Pareto optimal solution set
2. choice of the solution preferred.

Introducing the preference function P, to overcome the non-comparibility of objective functions in (7) the following is obtained:

$$\min_{U} P(f(U)),$$

$$g_i(U) \leqq 0, \qquad i = 1(1)\,n.$$

The preference function is defined for all values of the multidimensional objective function $f_i(u_i)$ and is used as a criterion to determine the rank order of the decision makers and to find the preferred solution within the set of the Pareto-optimum solutions. According to HWANG, PAIDY and YCON (1980) three methods are distinguished to solve the poly-optimization of a multi-criteria control problem: the objective method, the compromise method and the reference point method.

Our present approach to solve multicriteria problems of water quality control is based upon PARETO-optimization methods placed at user's disposal in an interactive manner by the program system REH (STRAUBEL und WITTMÜSS, 1986). In spite of the fact that no special requirements for model equations and goal functions have to be satisfied in using algorithms of REH the full version of SALMO will be implemented in the software environment of REH.

Provided that the first tests will be successfull, an extension of optimization problems towards a more regional view is planned where compromises have to be found for the different goal functions of water systems and users of drainage areas like agriculture, forestry and industries.

7. Eutrophication Control of the Eibenstock Reservoir. A Case Study

In the following, the concept of model-based water quality control developed in Chapter 6 will be applied for objectifying eutrophication control of the Eibenstock Reservoir.

The Eibenstock Reservoir was placed into operation in 1980 in order to supply Karl-Marx-Stadt, Zwickau as well as other cities and communities with drinking water. The damming phase was finished in 1985. With a volume of 71.5 millions m^3 and a catchment area of 200 km^2 it belongs to the biggest reservoirs of the GDR. The catchment area is characterized both by industrial, agricultural, forestry and also urban utilization and can be subdivided in the following way:

— 80% continuously wooded area
— 15% cultivated area with 11 communities, several industrial plants, live-stock farms, roads and railway lines
— 5% land in cultivation, mainly in the form of pastures.

The orthophosphate load of the reservoir resulting from specific uses of the catchment area can be differentiated as follows according to BENNDORF, UHLMANN and PÜTZ (1981):

— 48% sewage and runoff from communities
— 35% sewage from industrial plants
— 9% natural geochemical load
— 6% waste water from live-stock farms
— 2% runoff and leaching from agricultural areas.

The total orthophosphate load amounts to 1.45 g/m^2 y. This value is far beyond the crucial limit from meso- to eutrophic conditions of 0.65 g/m^2 y as calculated according to the P-load concept by Vollenweider (BENNDORF, 1979a). Consequently, in 1981 a comprehensive programme for managing the catchment area of the Eibenstock Reservoir was initiated, in the framework of which a variety of measures has already realized. Thus, one focused on external preventive control of the reservoir hitherto covering the following measures:

— Commissioning of five-pre-dams for the purpose of P-elimination
— Commissioning of a waste water by-pass line
— Reconstruction of a waste water plant
— Evacuation, reconstruction or shut-down of live-stock farms
— Stopping usage of pastures

The programme envisages the implementation of further measures: for external preventive control of the reservoir by 1990. Using the concept of model-based water quality control it should be checked how far the programme for managing the catchment area consequently executed will warrant a reliable oligotrophication of the Eibenstock Reservoir and which alternative or complementary control policies, respectively, could be taken into account in future with regard to measures for internal preventive control of the water. The input file for a mean year of the Eibenstock Reservoir with assumed complete damming according to KRUSPE (1984) served as a basis of investigations.

7.1. Scenario analysis

The reference run of the scenario analysis for the Eibenstock Reservoir was based on the estimated external orthophosphate load without regarding the elimination performance of pre-dams and restauration measures within the catchment area. According to the decision table in Tab. 6 four scenarios were defined. Scenario 1 served for simulating the

Table 6: Decision table of scenario analysis for the Eibenstock Reservoir

		Control measures			
		Managing the catchment area	Phosphorus elimination plant	Artificial destratification	Manipulation of food-web
Problem definition: management target	Decrease in external phosphorus load	×	×		
	Decrease in secondary load by eutrophication	×	×	×	×
	Improvement of oxygen balance	×	×	×	×
Scenario specification	Scenario 1	×			
	Scenario 2		×		
	Scenario 3			×	
	Scenario 4			×	×
Scenario control by the model	Decrease in nutrients concentration of water inflow	×	×		
	Increase of mixing depth in summer			×	
	Decrease in water temperature in summer			×	
	Decrease in mortality of zooplankton				×

system behaviour of the Eibenstock Reservoir where the complete realization of the programme for managing the catchment area in 1990 was assumed. The nutrient concentrations in the inflow of the reservoir used as a basis were calculated corresponding to the theoretically expected nutrient elimination performance of the measures planned and already executed, respectively (KRUSPE, 1984). In scenario 2 a decrease in the external orthophosphate load by 90% was defined using the Wahnbach-method (BERNHARD and CLASEN, 1981). Thereby, it had to be checked if the Wahnbach-method showed an alternative to constraints of agricultural uses in the catchment area of a reservoir in the sense of an extension of the pre-dam concept by technologies for physical and chemical waste water treatment, as expressed by BERNHARD (1981).

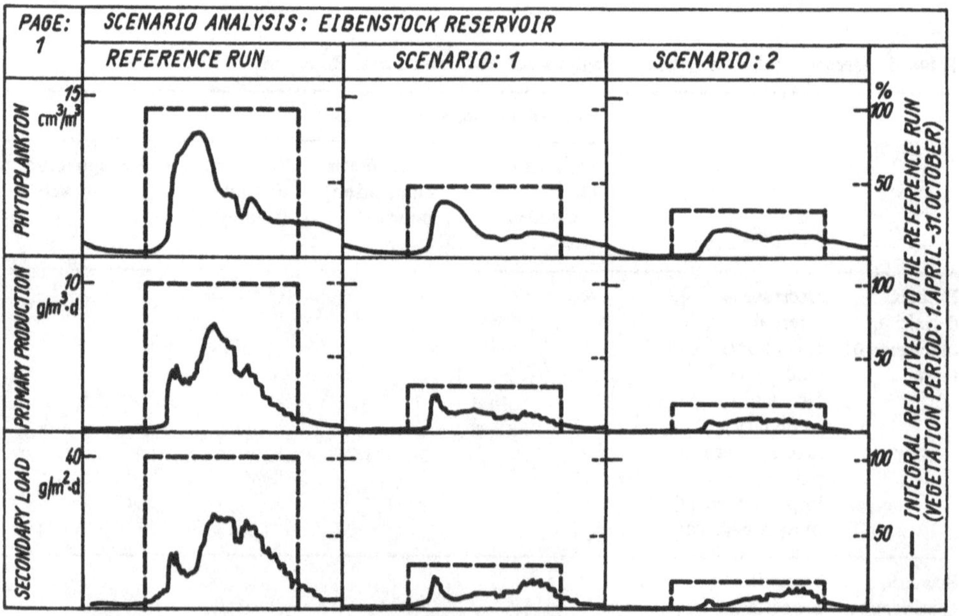

Fig. 51: Scenario analysis for the Eibenstock Reservoir.

Scenario 1: Extensive policy for restauration and management of the catchment area

Scenario 2: Treatment of water inflow by the Wahnbach-method for phosphorus elimination up to 10%

Scenario 3: Artificial destratification of the water body during summer

Scenario 4: Artificial destratification of the water body during summer and manipulation of food web by predatory fish for reducing the mortality of herbivorous zooplankton up to 1%

Scenario evaluation by means of comparing the trajectories and relative integrals within the vegetation period with the reference run.

Page 1 & 5: Trajectories and integrals of the system quantities phytoplankton, primary production and secondary load

Page 2 & 6: Trajectories and integrals of the system quantities coefficient of extinction, zooplankton and mortality of zooplankton

Page 3 & 7: Trajectories and integrals of the system quantities detritus and hypolimnic oxygen

Page 4 & 9: Trajectories and integrals of the system quantities dissolved orthophosphate, inorganic nitrogen and N:P ratio

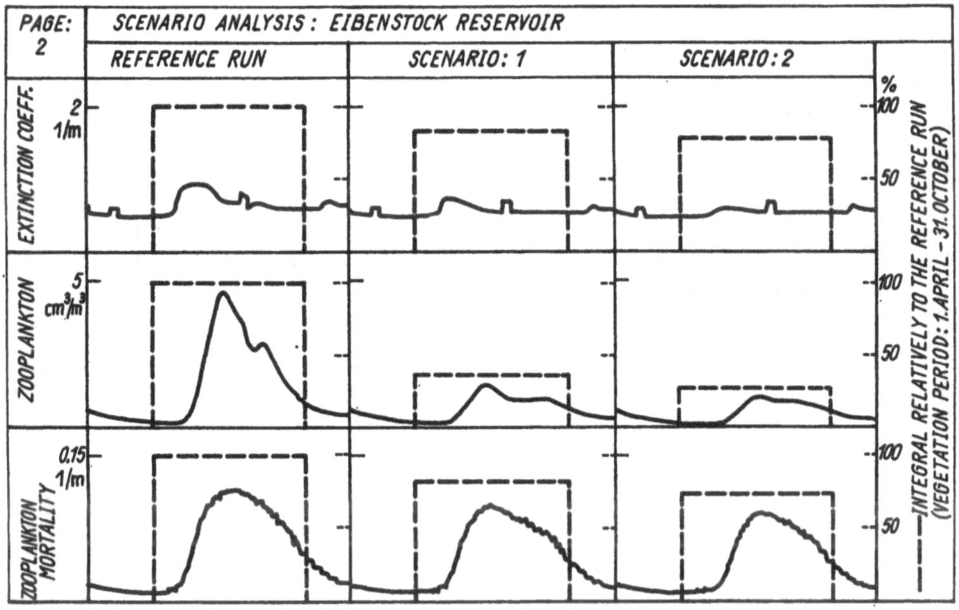

Fig. 51, Page 2

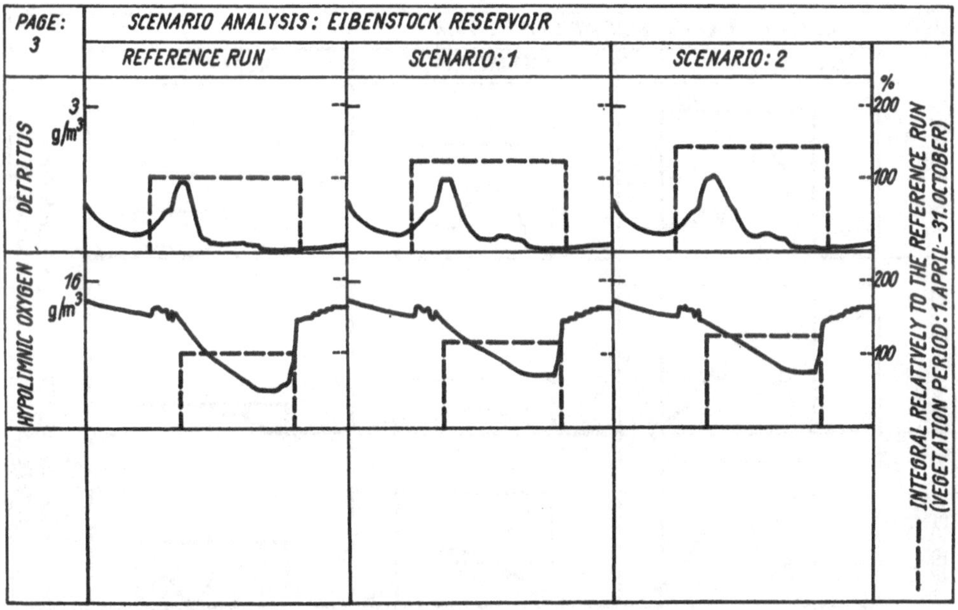

Fig. 51, Page 3

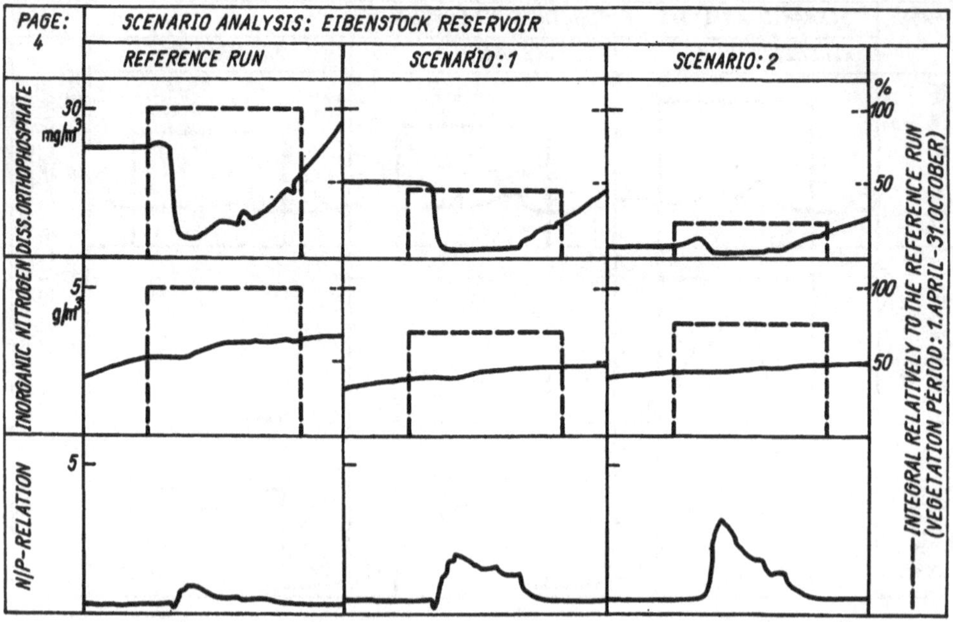

Fig. 51, Page 4

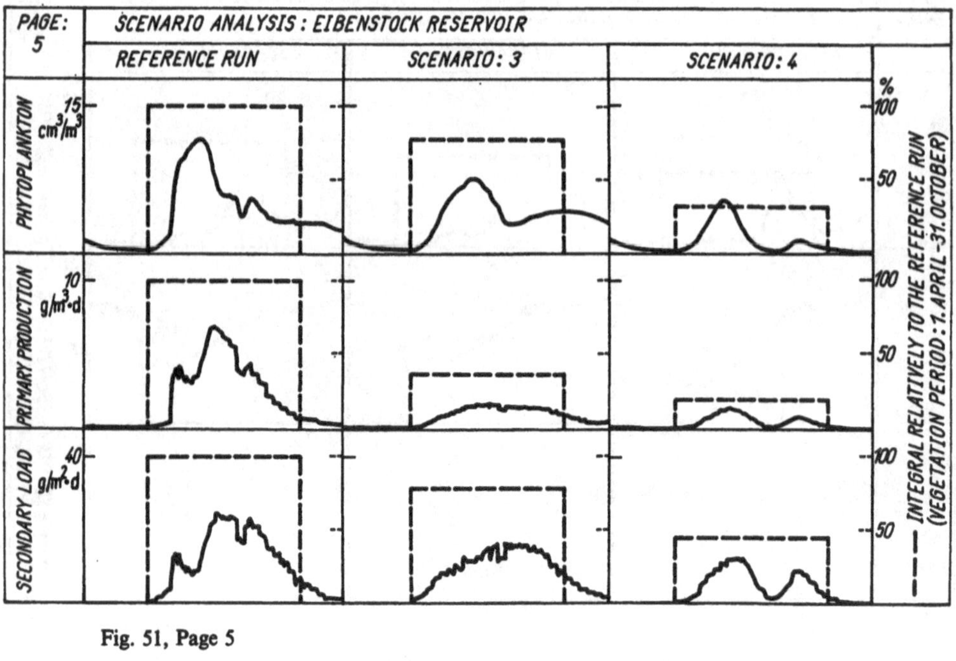

Fig. 51, Page 5

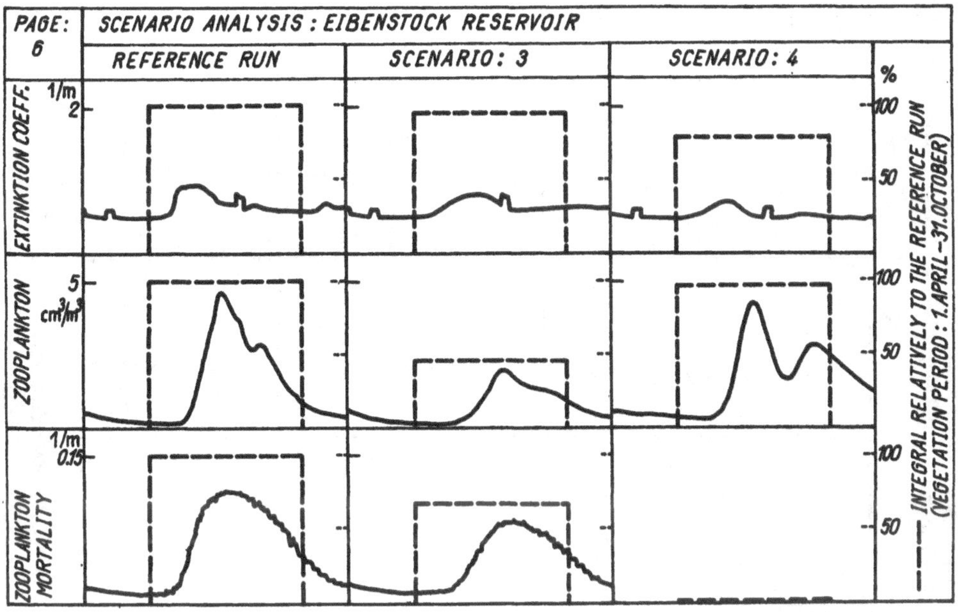

Fig. 51, Page 6

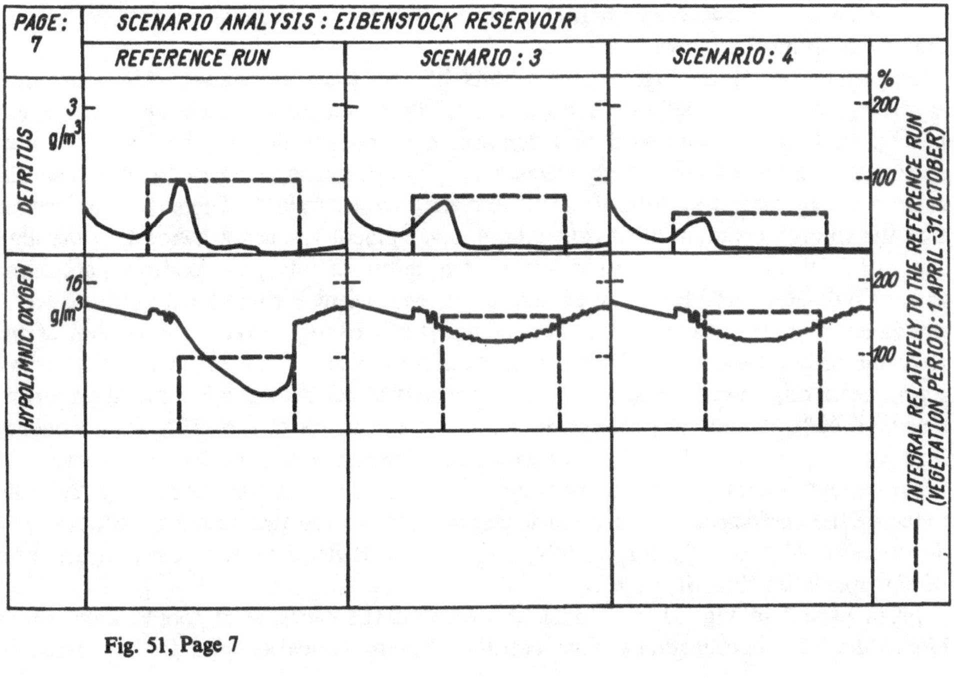

Fig. 51, Page 7

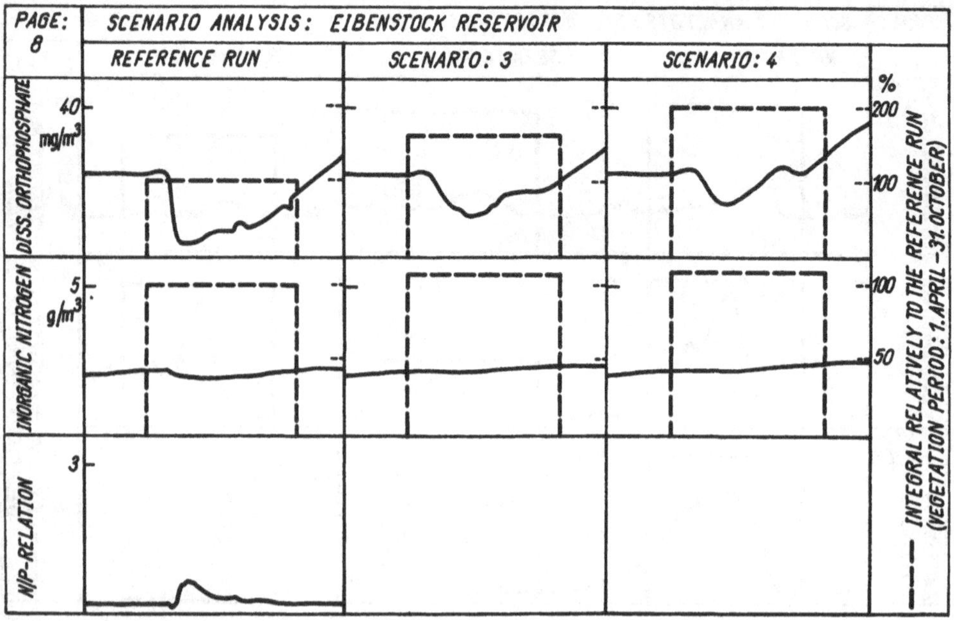

Fig. 51, Page 8

Scenario 3 was defined for simulating the effects of artificial destratification of the water body according to Klapper (1977). Finally, it had to be clarified with the help of scenario 4 to what extent eutrophication control of the Eibenstock Reservoir would be promissing using exclusively internal control measures of artificial destratification and manipulation the food-web.

In pages 1 to 8 of Fig. 51 the results of the scenario analysis are represented according to the size defined in Section 6.1. By means of the trajectories in page 1 of Fig. 51 it can be estimated that conception of restauration of the catchment area until 1990 can effect a marked decrease of phytoplankton biomass to 50% and the corresponding secondary load to 25% according to scenario 1. Thereby, it is striking that the spring maximum of phytoplankton development reaches values within the limit range of 5 cm^3/m^3 indicating the transition from meso- to eutrophic conditions according to the GDR-Standard. This may be due to the insufficient P-elimination performance of pre-dams in spring caused by too low temperatures owing to the altitude level of the reservoir (538 m above sea-level). By means of the Wahnbach-method according to scenario 2 a permanent oligotrophication of the reservoir could be achieved at a P-elimination performance of 90%, thereby phytoplankton biomass could be reduced to 35% and secondary load to 15% compared to the reference run. Approximately to the same extent an improvement of transparency of the water can be achieved through scenarios 1 and 2 by reducing the coefficient of extinction to about 80%. In the two scenarios, zooplankton decreases to 40 and 30%, respectively, because of a reduced nutrient basis in the form of phytoplankton (Fig. 51, page 2).

From page 3 in Fig. 51 it becomes evident that the decrease of zooplankton causes higher detritus concentrations since detritus likewise serves as food for zooplankton.

Due to the lower secondary load of water as a result of scenarios 1 and 2 as mentioned above an improved hypolimnic oxygen concentration to 110 and 125%, respectively, is reached.

In page 4 of Fig. 51 the changed nutrient conditions for the simulated scenarios 1 and 2 are displayed. Implementation of the management conception within the catchment area according to scenario 1 results in a decreased orthophosphate concentration to 50% and of nitrogen concentration to 75%. According to scenario 2, the internal orthophosphate concentration could be reduced to 25% and nitrogen concentration to 75% by applying the Wahnbach method. Finally, the annual trajectories of N:P ratios are plotted in the lower third of page 4, Fig. 51. As to be expected, the dominating P-limitation of phytoplankton growth during the vegetation period is indicated.

Scenarios 3 and 4 can be evaluated on the basis of pages 5 to 8 of Fig. 51. From page 5 it can be seen that artificial destratification of the water according to scenario 3 causes phytoplankton development to be decelerated. Consequently, biomass concentration and secondary load can be reduced by 25% roughly. Obviously, the heavily reduced primary production (40%) because of the lower specific light consumption of phytoplankton is compensated due to diminished pressure of zooplankton grazing, which can be explained by lower mean water temperatures. The combination of artificial destratification with measures in terms of biomanipulation according scenario 4 turns out more effective. For biomanipulation a decreased zooplankton mortality to 1% influenced by an increased predatory fish stock is defined. By means of scenario 4 a significant decrease in both phytoplankton biomass to 30% and secondary load to 50% at equally high external nutrient load is achieved.

Page 6 of Fig. 51 shows that no-perceptible improvement of transparency of water can be obtained by scenario 3, whereas transparency can be increased using scenario 4 when diminishing the coefficient of extinction by 20%. Zooplankton development is seriously delayed by artificial destratification, while a combination of destratification and biomanipulation according to scenario 4 raises zooplankton biomass to the level of the reference run.

Page 7 of Fig. 51 indicates that the detritus concentration within the whole water body is lower by 20% roughly after successful artificial destratification than at the consideration of epilimnion during thermal stratification in summer. Biomanipulation effects a clear decline of detritus concentration to 50% due to enhanced pressure of zooplankton grazing. In the two scenarios 3 and 4, destratification results in a significant improvement of oxygen conditions in water.

Page 8 of Fig. 51 reveals that the unilateral internal eutrophication control always effects an increase in nutrient concentration in the water such that there is a potential eutrophication danger in case of an internal control failure. Additionally, the probability of internal nutrient loads owing to progressing nutrient accumulation in the sediment of water is increasing over the years. In future, statements concerning these processes will require separate balancing of nutrient conditions in sediment. Currently, the conditions are created by developing a sediment submodel for the model SALMO. As a result of scenario analysis of the reservoir Eibenstock it can be estimated that an efficient preventive control of eutrophication will be possible for the catchment area of the reservoir — represented in scenario 1 — by means of the restauration and management programme, however, there is no guarantee provided for a permanent oligotrophication of the reservoir. Due to insufficient nutrient elimination performance of pre-dams in spring, algae mass developments in the reservoir cannot be excluded, thus it represents a risk factor concerning the success of control policy. The Wahnbach method — plotted in scenario 2 — has proved to be an alternative policy to external preventive control of eutrophication. Obviously, by means of the Wahnbach method a compromise is possible between a certain degree of socially required usages of the catchment area, e.g. by agriculture, and the guarantee of a reliable oligotrophication. Internal preventive control via artificial destratification and biomanipulation have turned out to be promising control policies, particularly in combination (scenario 4). At unilateral control of these measures during a longer time period the internal nutrient load of sediment will become a risk factor for a permanently successful oligotrophication of the water due to progressing nutrient accumulation in sediment. So far, external preventive control is to be attributed highest priority, with its completion by internal control measures being desirable.

7.2. Stability analysis

In the following, structural stability of ecosystem behaviour of the Eibenstock Reservoir at a combination of control of external orthophosphate load and artificial destratification will be investigated. In Figs. 52 and 53, the isocline surfaces of the state variables phytoplankton and zooplankton and the associated projections of isolines into the control space are represented according to the size agreed upon in Section 6.2. The control space is defined above the measures of artificial destratification (X-axis) and change of external orthophosphate load (Z-axis). By means of artificial destratification a successive extension of the completely mixed water body in spring up to the beginning of autumn is controlled in decade intervals. The change of external orthophosphate load is regarded in relation to the reference year in the range of 10% to 200%. Thereby, orthophosphate load in the reference year is considered as 100%. In analogy to Section 7.1, a mean year of the Eibenstock Reservoir, is used as a reference year without regard to management and restauration measures in the catchment area. The projections of isolines into the control space refer in Fig. 52b to a biomass gradient of phytoplankton of 1 cm^3/m^3 and in Fig. 53b to a biomass gradient of zooplankton of 0.05 cm^3/m^3.

The isocline surface in Fig. 52a demonstrates that phytoplankton development will rise erratically if artificial destratification is initiated for a period of ten days — starting at the end of the spring period. The slope gradient is growing with increasing external nutrient load. Evidently, the reason for this can be seen in optimum light and temperature conditions for phytoplankton growth. Zooplankton profits from these conditions as well, both directly through water temperature and indirectly through an improved food supply in the form of phytoplankton biomass (see Fig. 53a). Then, phytoplankton will be controlled effectively from a minimum destratification period of ten decades. Until that point, phytoplankton growth will be linear, approximately, with increasing external nutrient load. As to be seen from Fig. 53a, zooplankton growth will be delayed primarily within these ten decades due to deteriorating temperature conditions resulting in a compensation of reduced primary production of phytoplankton. A structurally stable region for supporting the most important control function of zooplankton at eutrophication control of the Eibenstock Reservoir will be obtained if the completely mixed water body in

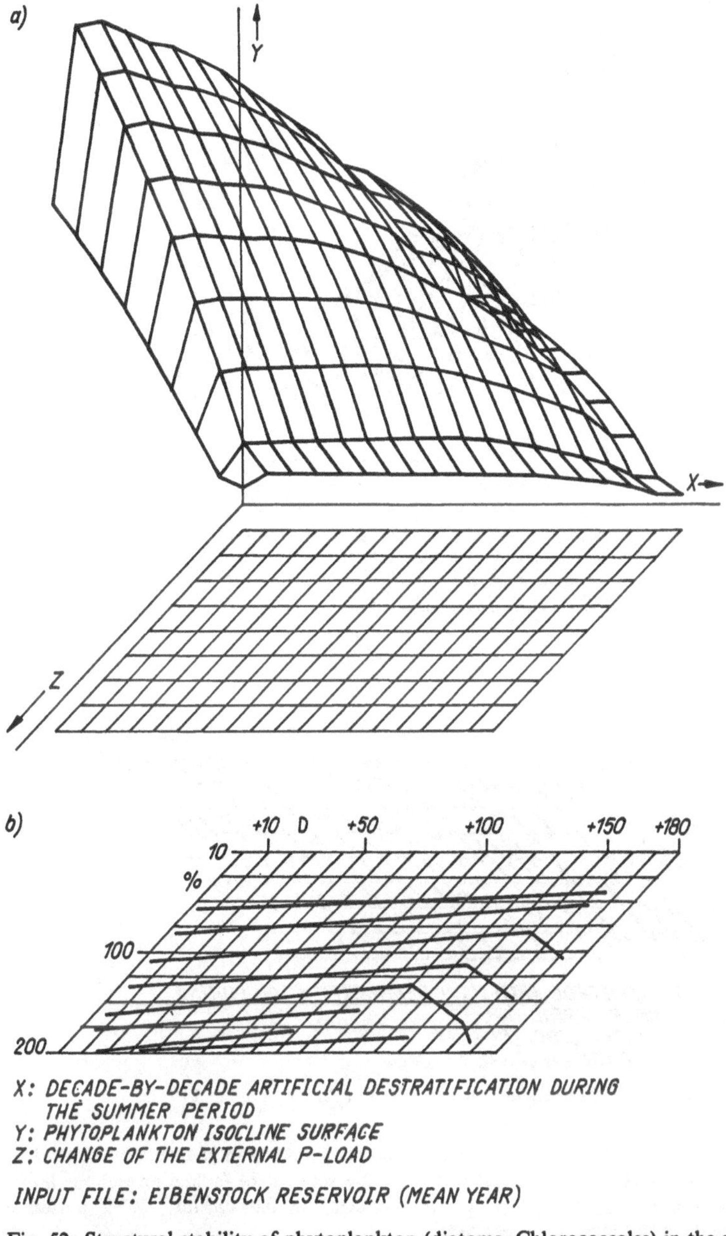

Fig. 52: Structural stability of phytoplankton (diatoms, Chlorococcales) in the relevant control space of the Eibenstock Reservoir. a) Phytoplankton isocline surface, b) Isoline projections into the control space. Isolines correspond to biomass gradients of 2 cm^3/m^3; 100% corresponds to the nominal control level in a mean year

spring is prolonged by 10 to 16 decades as a maximum by means of artificial destratification (see Fig. 53b). The stability of this region will be maintained even at increased external orthophosphat load to about 190%.

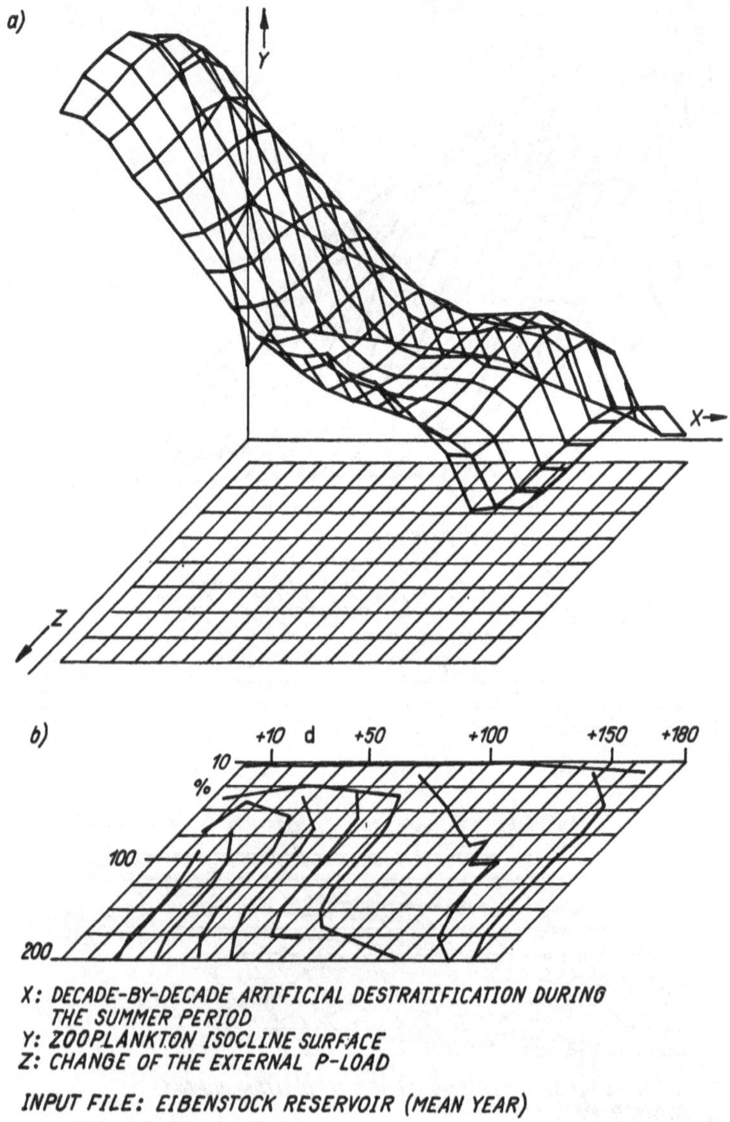

Fig. 53: Structural stability of herbivorous zooplankton in the relevant control space of the Eibenstock Reservoir. a) Zooplankton isocline surface, b) Isoline projections into the control space. Isolines correspond to biomass gradients of 0.05 cm^3/m^3; 100% corresponds to the nominal control level in a mean year

7.3. Dynamic cost optimization

In a first test phase the program system for scalar optimization with state phase constraint was applied to the Eibenstock Reservoir according to Fig. 50 for the optimization of dynamic control of external orthophosphate load using the Wahnbach method. In order to keep a justifiable computing time the corresponding control variable

u_w was discretized into six equidistant control spans within the control period of one year. As the lower limit of control by u_w 0% was defined and as the upper limit 90% according to the maximum elimination performance of the Wahnbach-method so that the control constraints will be as follows:

$$0 \leqq u_{wi} \leqq 0.9\,, \qquad i = 1(1)\,6\,.$$

The initial values for u_{wi} were roughly estimated at first and precisized successively according to the observed system sensitivity. State space constraints were defined as follows:

$$z_2(t) + z_3(t) \leqq 1$$

$$\text{for} \quad 0 \leqq t < 90\,, \qquad 270 \leqq t < 360\,.$$

and

$$z_2(t) + z_3(t) \leqq 5$$

$$\text{for} \quad 90 \leqq t < 270$$

$$\text{with} \quad \Delta t = 1d\,.$$

As initial values for the Lagrange multipliers

$$\lambda_j = 10^6\,, \qquad j = 1(1)\,360$$

and for the penalty factor $r = 0.1$ were selected.

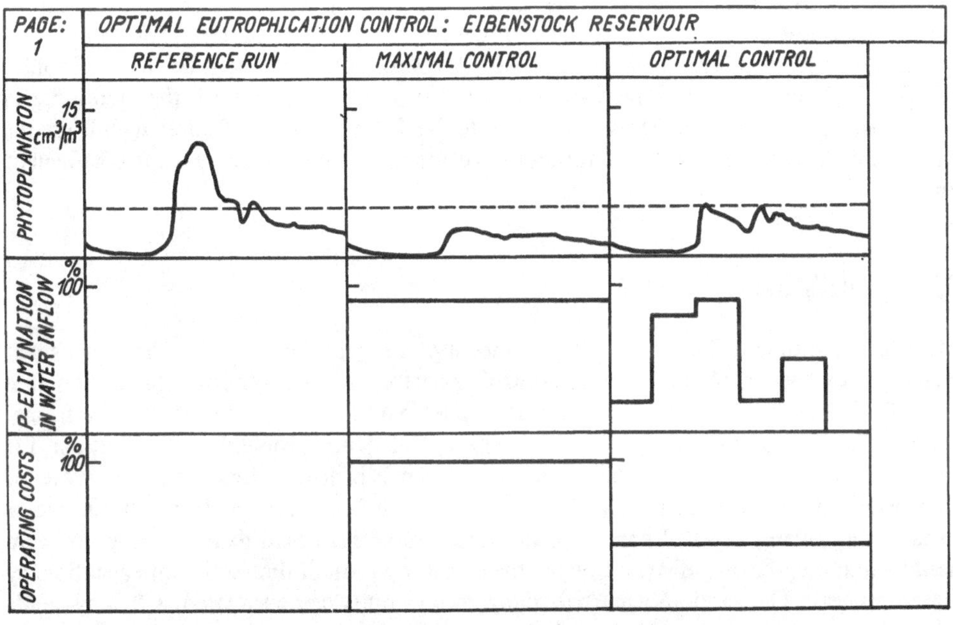

Fig. 54.1: Optimal eutrophication control of the Eibenstock Reservoir by a phosphorus elimination plant.

State trajectories of phytoplankton (figures above), control trajectories for the phosphorous elimination (figures in the middle) and annual relative operating costs (figures below) for the uncontrolled case (left-hand figures), the maximal controlled case (figures in the middle) and the optimal controlled case (right-hand figures).

The control time of one year was divided in six equidistant control periods of two months

By an optimization run at the ROBOTRON computer 1022 the result represented in Fig. 54.1 could be obtained for the reservoir Eibenstock Reservoir with a CPU-time of ten hours. In Fig. 54.1 state trajectories for the total biomass of phytoplankton, control trajectories for reducing the external nutrient load and the trajectories of relative operating costs are plotted for comparative purposes of three cases: uncontrolled case with $u_w = 0$, maximally controlled case with $u_w = 0.9$ and optimally controlled case. The operating costs OC_w presented were computed in relation to operating costs of 100% for the case maximally controlled. At the state trajectories in Fig. 54.1 the limit of 5 cm^3/m^3 was drawn additionally as a dotted line in order to warrant meso- and oligotrophic conditions, resp., according to the GDR-Standard. Fig. 54.1 shows that both maintenance of the limit for phytoplankton development and clearly lower operating costs can be achieved by an optimum control of external orthophosphate load according to the Wahnbach method. At control intensities of about 20% for u_w in January—February as well as in July—August of the year, of about 80% in March—April, of about 90% in May—June, of about 40% in September—October and of 0% in November—December, phytoplankton biomass is permanently below 5 cm^3/m^3 and operating costs are reduced to 40% roughly.

The example of dynamic optimization of the Wahnbach method for reducing external nutrient load of the Eibenstock Reservoir indicates that in future technologies which can be controlled dynamically in the sense of preventative control will offer a chance to find compromises between certain uses of the catchment area by means of optimization.

First of all, the combination of the control variable u_w with the control variable u_D for artificial destratification is envisaged in the second phase of optimization in order to obtain a further decrease of operating costs. A condition for this will be the availability of „fast" algorithms. In a parallel manner, the implementation of the multicriteria optimization according to Section 6.3.2. is to be focused on. In future, it will enable us to achieve optimum, system-interlinking regional solutions which permit cost/benefit analyses.

7.4. Risk analysis

In order to estimate the probability of meeting the goal function for the Eibenstock Reservoir by means of the optimal control trajectory for phosphorus elimination (see Sect. 7.3.) PETZOLDT (1988) applied the model SALMO in a Monte Carlo framework. First step of the approach was to specify adequate probability distributions for basic driving variables within the input file of the Eibenstock Reservoir which are of stochastic nature. Such driving variables are water inflow, light intensity at the water surface and mixing depth. In the case of water inflow historical data of ten years were used to adapt log-normal distribution for the transformation of time series into a stationary Gauss-process. The randomly picked, daily inflow rates are correlated with any loads of nutrients and organic matter but also with water volume, water outflow rates and the turbidity factor of the light extinction term. Daily values of light intensity and mixing depth are generated randomly by using of autocorrelations for adapted uniform distributions of the bivariate Plackett-type. In both cases synthetical cover curves derived from measured time series have to be fulfilled. The generated mixing depths are correlated with the epi- and hypolimnic water volumes during summer.

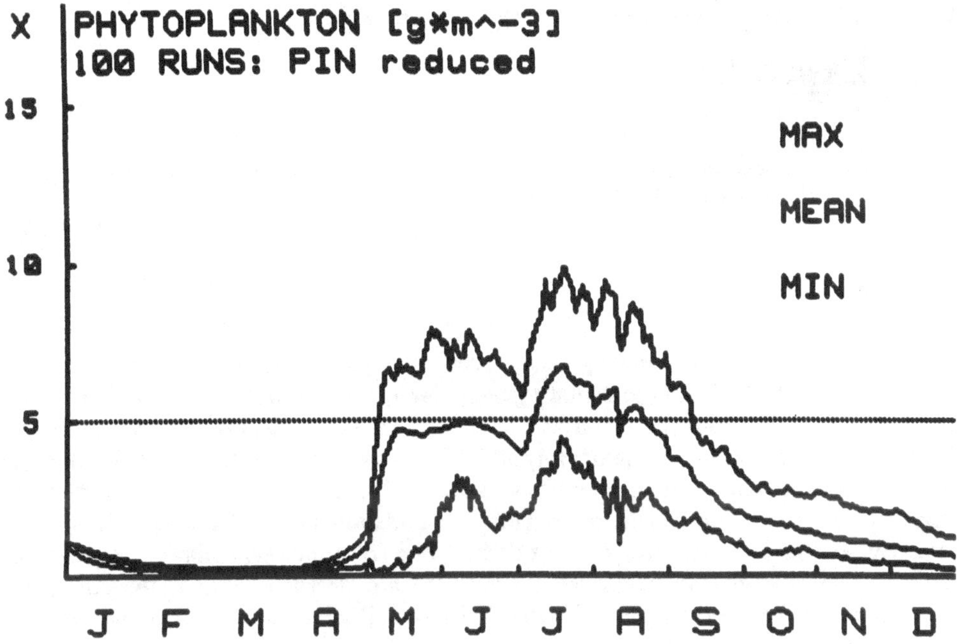

Fig. 54.2: Confidence interval of predicted phytoplankton biomass by Monte Carlo simulation with 100 runs of model SALMO (——— limit value for oligotrophic conditions as defined in the TGL 27885/01). After PETZOLDT (1988).

In Fig. 54.2 results of a Monte Carlo simulation with 100 runs of SALMO are represented, where the above mentioned stochastic inputs were used simultaneously. By the hand of the confidence interval of the predicted phytoplankton biomasses which is limited by the synthetical minimum and maximum trajectories the following conclusions can be drawn. The threshold value for oligotrophic conditions of 5 cm^3/m^3 phytoplankton biomass can be violated from May up to the beginning of September where the threshold value for mesotrophic conditions of 10 cm^3/m^3 will be maintained stably. The synthesized mean trajectory permits to assume that a more than 50% probability for violating the aimed limit value of 5 cm^3/m^3 only occurs from July 5th up to August 20th. To reduce the risk of such events the phosphorus elimination rate of 20% during July and August as proposed in Sect. 7.3. should be increased up to about 50%.

8. Further Developments Towards Expert Systems

Expert systems are programs or sets of programs designed to multiply the value of real experts by putting their knowledge at the fingertips of nonexperts.

A. Lampert

The model-based decision making using predictive models is a practicable way to objectify the control of environmental systems. As it was shown, this method is based on precise scientific knowledge on the composition and structure of such systems. However, models are always homomorphies of real systems, i.e. they provide „model truths" a priori, which are incompletely, simplified and aggregated. A *first* step to overcome this restriction should be to build a simulation library for a definite system class, e.g. for lake ecosystems. In future, the users of simulation libraries should be enabled to include alternative model information into decision making, the reliability of which can be assessed by the computer. Thereby, all advantages of computer and of optimization should be exhausted.

A *second* important step to further objectify the control of environmental systems is the development of system specific data bases which are fed by actual, historical and synthetical data. In obtaining actual data, it is more and more the question to achieve the required process nearness by using of reliable measurement and telemetry technique. As not all the variables of interest of the system can be measured, it will be necessary to define appropriate indicator or „surrogate" variables. This task should be brought into line with the demand for an optimum design and operating mode of such monitoring systems. By real-time data, the possibility of on-line control is given provided that appropriate models and computers for real-time simulation are at disposal including digital filter algorithms. Within off-line control, data bases considerably contribute to the completion and validation of model information towards Kalman's canonical system.

Decision making for environmental systems is characterized by a high extent of uncertainty and risk factors caused by the complexity, nonlinear dynamics and natural stochasticity of such systems. There was progress in using of simulation systems to assess decisions or policies under uncertainty by deterministic and probabilistic methods, e.g. sensitivity analysis, Monte Carlo simulation, constrained optimization under uncertainty. But in spite of the fact of analytical uncertainties which are inherent models and parameters in a *third* step it proves to be of a considerable value to make available additionally expert knowledge to decision makers. Meanswhile methods were developed in the field of AI-research enabling a more intuitive handling of decisions or policies. Empirical decision rules, expert judgements, heuristics, facts and special knowledge of several tangent disciplines can be acquired, accumulated and placed at any user's disposal by knowledge engineering.

The so-called expert systems, which can be described as result of knowledge engineering in a narrow domain, will open a new quality of computer-based decision making for environmental systems. Architecture and operation of expert systems in an optimal manner promise to permit an integrated utilization of the three information columns for decision making mentioned above: simulation library, data base and knowledge base.

An approach is proposed which combines AI- and simulation methods to render decision making in the field of operative control, management and planning water quality of lakes and reservoirs more „objective". The expert system as used constitutes an elaborated version of the simulation system which is based on the model SALMO. In this context, the simulation system is handled as a main component of the knowledge base of the expert system. A parallel coupling of both systems provides an adequate software architecture in combining the FORTRAN 77-language and the AI-language PROLOG.

8.1. Expert systems

Expert systems are computer programs designed to simulate the problemsolving behaviour of a human who is an expert in a narrow domain (DENNING, 1986). They dispose of reasoning power which depends critically on facts and heuristics as used by experts. The credibility of an expert system" is greatly enhanced if it can explain to the user the rationale that led to the findings of expertises. Users interact with expert systems in a consulting dialogue.

According to HAYES-ROTH, WATERMAN and LENAT (1983) the general architecture of expert systems consists of three parts: an interactive user interface, a control and inference structure, and a data and knowledge base (see Fig. 55). The *interactive user interface* permits the dialogue between user and system. In natural language interface systems a so-

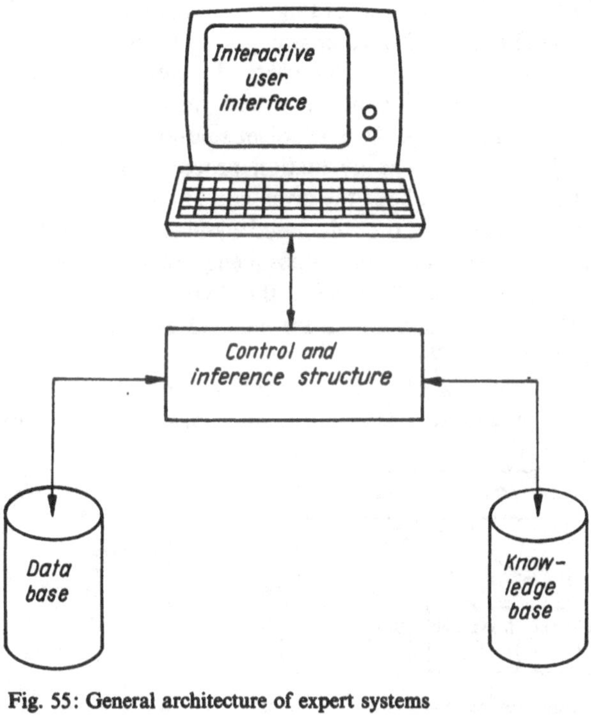

Fig. 55: General architecture of expert systems

called parser is available. This is a processor that converts the user's input into statements the expert system can understand. Such parsers, however, require large amounts of data storage capacity for their dictionaries and their grammatical algorithms. Therefore, most expert systems presently employ a user interface that permits the user either to enter specific commands or to select menu options. The commands will then be transferred to the control and inference structure, which provides a mechanism for interpreting commands, defining the problem solving approach and access to the data and knowledge base. The *control and inference structure* is a set of rules or algorithms that governs the ability of the system to draw conclusions. The *data base* includes declarative knowledge in the form of data or facts about the particular problem domain, where as the *knowledge base* contains procedural or relational knowledge in the form of procedures, rules and models for combining data. The data and knowledge base are generally collected and compiled by experts in the relevant subject areas. Expert systems which a priori are placed at a user's disposal with empty data and knowledge bases are named shells.

As stated by different authors (e.g. FEIGENBAUM, 1984) knowledge acquisition and representation prove to be a bottle neck in elaborating expert systems. Generally, three types of systems for representing knowledge can be distinguished (see Denning, 1986): logic programming systems, rule-based systems and frame-based systems.

Logic programming systems use predicate calculus for representing declarative facts or procedural statements. E.g. a particular predicate statement using the language PROLOG like $A:-B_1, B_2$ can be interpreted declaratively that A is true if B_1 and B_2 are true and as a procedure for producing a state satisfying A by executing B_1 and B_2.

Rule-based systems are structured hierarchically by using of rule sets. Each rule set comprises a series of rules of inference according to the pattern „if conditions then consequences". Forward chaining or backward chaining are usually used for evaluating consequences of the initial conditions given by users. In case of forward chaining problem solution starts from the initial conditions and moves toward the ultimate consequences. If backward chaining is used for a problem solution the desired conclusion or goal state is known, hypothetically. The evaluation proceeds by tracing rules until the ultimate antecedents of a hypothesis about the given conditions have been found.

The most suitable form of representation rule-based systems is an AND/OR-graph (see Fig. 56) in which node A represents the consequence of conditions represented by the nodes $B_1 \dots B_n$. A circular arc indicates the logic conjunction (AND) of the enclosed conditions, otherwise the conditions are handled as alternatives to fulfill A in the sense of the logic disjunction (OR).

A *frame-based system* consists of a hierarchy of descriptions of objects referred to in the rules, where a frame means the description of a class of objects or a single

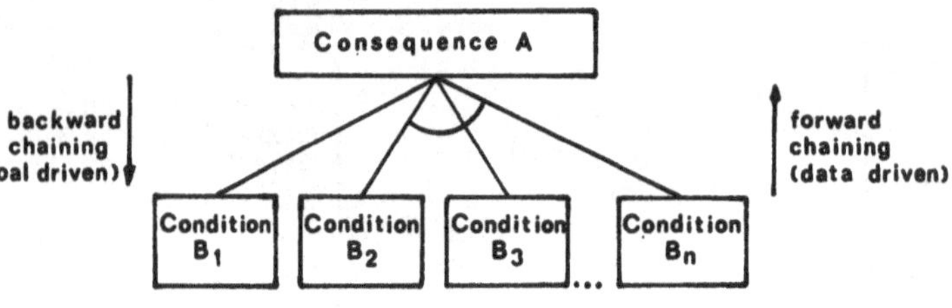

Fig. 56: AND/OR-graph for the logic rule A if B_1 or B_2 and $B_3 \dots$ and B_n

object by a collection of facts and data about the objects. The hierarchy of frames is defined by relations, and functional relationships among frames are described by rules. The advantage of this system type consists in the ability to identify properties of a specific object in a direct way by locating it in the hierarchy of frames.

The languages mainly used for knowledge representing and processing in expert systems are LISP (LISt Processing) and PROLOG (PROgramming in LOGic).

LISP (WINSTON and HORN, 1981) is a very small language which employs the same representation for instructions as for data. The basic LISP data elements are called atoms and lists. Atoms can be numbers, strings or symbols. A list represents a binary tree node with two pointers which can refer to other lists or atoms. There is an operator to create new nodes; unreachable nodes and atoms are garbage collected. Symbols have associated property lists where doublets of the form (property_name property_value) are kept. Standard functions allow properties to be consulted, updated or removed. Some properties are used by the language interpreter in a standard way; e.g. the symbol value and the function definition if the symbol has been defined to be a function identifier.

PROLOG (CLOCKSIN and MELLISH, 1981) uses the notation of predicate logic and contains no imperative statements, only assertations and rules of deduction. PROLOG computation is couched in terms of requests for proof of a query. Usually the query contains variables and, as part of the proof process, PROLOG will find variable bindings that make the query true. Although initially designed for natural language handling, PROLOG has a much wider range of application and has been seen as a replacement for relational data base systems. PROLOG statements can be considered to have the following general form:

⟨left part⟩ :— ⟨right part⟩

with the meaning that the ⟨right part⟩ implies the ⟨left part⟩. If the right part is missing the statement is interpreted as asserting a fact.

Both languages dispose of interpretive power, which enables programs that can manipulate different types of data as well as other programs, and to change the running program's behaviour. Moreover, PROLOG is well suited to parallel search and is, therefore, predestinated for future powerful computers incorporating parallel processing.

Depending on the quality of their knowledge representation expert systems can be subdivided into shallow and deep systems (DENNING, 1986). The validity of *shallow systems* is limited because their knowledge bases include more facts than rules. Consequently, the proofs of their conclusions are short and they often fail in the attempt to be applied to problems for which they have not been tested. By contrast *deep systems* are based on models of phenomena in their domains. The proofs of their conclusions are usually long. Along with backward chaining so-called deep systems should be able to generate hypotheses about initial conditions; in the course of forward chaining they should be able to explain discrepancies between predicted and observed final states by generating hypotheses concerning the components affected.

The characteristics of deep expert systems provoke associations with customary traditional simulation systems. For this reason, some authors of the simulation scene dealt with conditions for developing simulation systems into deep expert systems. The most essential differences between them as emphasized by SHANNON, MAYER and ADELSBERGER (1985) can be summarized as follows:

1. Simulation systems executive the statements of their algorithms in sequential order. Expert systems dispose of the ability to find automatically the model which fulfils the desired specifications and to execute an appropriate approach to obtain the desired solution.
2. Expert systems gain the ability mentioned above by including symbolic data of heuristic knowledge. Simulation systems are driven primarily by numerical data.
3. Simulation systems base on integrated algorithms which perform data processing and simulation control. In expert systems control structure, data and knowledge base are separated, with each of them being modifiable easily without affecting the others.

Thus a traditional simulation system is transformed into an expert system, once most of the decisions that are now made by the simulation expert are built into the system, successively.

At present, the main application areas of real existing expert systems are business, fault diagnosis in computer industry, diagnosis in medicine (MICHAELSEN and MICHIE, 1983), assessment of chemical danger to the environment and community health (HUSHON, 1986), genetic engineering (STEFIK et al., 1982) and in the water systems management (JOWITT and LUMBERS, 1985; JOWITT and JENKINS, 1986; CERTES, HUBERT and LETOLLE, 1986).

8.2. A deep expert system for water quality control

At present, an expert system is under construction according to the structure in Fig. 57. As of the beginning, it was intended to elaborate a deep expert system by incorporating advanced decision making software on the base of the simulation model SALMO. According to the taxonomy of O'KEEFE (1986), the objective is pursued by a parallel coupling of the expert system and the simulation system, the two interacting in a parallel manner (see Fig. 57). This approach promises to increase the credibility of the expert system by utilizing trajectory simulation of optional key process in the waters. In view of this structure, the expert system and the simulation system exist as separate software units implemented by PROLOG and FORTRAN 77 respectively.

The expert system consists of a dialogue module, a control and inference module and two sources of information: a knowledge base and a data base. In addition, the coupled simulation system acts as an excellent storage of human knowledge on the narrow domain of lake ecosystem's behaviour, provided that a predictive model like SALMO is available. In order to increase the credibility of the simulation system further it is planned to make available alternative predictive lake models — not only ecological but also hydrodynamic — along the lines of a simulation library as proposed by SWARTZMAN (1979).

The dialogue module is designed as a user interface that permits the user to communicate with the system by commands and menu options in an interactive manner. The module involved subroutines for including peripheral terminals, printers and plotters to monitor the problem solving process and to document results and conclusions in a userfriendly design. For these purposes the window technique as available in PROLOG proves to be advantageous. The control and inference module disposes of subroutines that in compliance with user's request compile the menu options. For this purpose it makes use of the data base, knowledge base and the simulation system. Assessments, judgements and conclusions

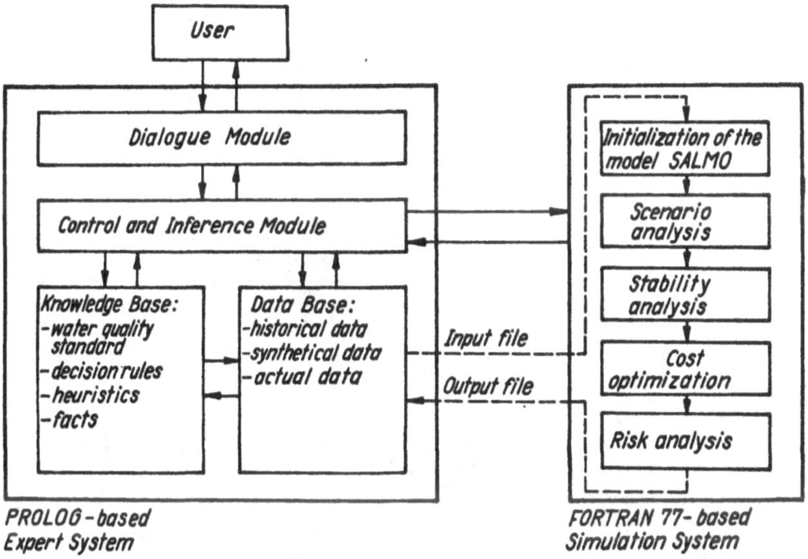

Fig. 57: Structure of the conceptional expert system for water quality control

generated by this module are always aimed at rational alternatives which are optional for the user.

In developing the knowledge base about 80 decision rules of the water quality standard of the GDR (TGL 27885/01) were coded by PROLOG. In Fig. 58 the AND/OR-graph of this module is represented. It allows to classify automatically any water under consideration into five water quality classes which are interrelated with requirements to raw water treatment for envisaged uses of the waters. The water quality index can be calculated using measured quality criteria of specific waters or with the help of synthetical data generated by the simulation system for the purpose of evaluating water quality predictions. Moreover, the knowledge base disposes of heuristics, facts and decision rules regarding damages which are subject to operational control. Such damages are e.g. algal blooms or external contaminations which have to be counteracted within hours or few days. In Figs. 59 and 60 segments of the rule-based module for operational control of algal blooms are represented, which are data driven in the sense of forward chaining. The underlying rules widely correspond to KLAPPER (1983).

The data base at present contains historical and synthetical data. In obtaining and including actual data, the question of availability of reliable measurement and telemetry systems is decisively. Such systems are presently not guaranteed. The historical data are annual records of water quality measurements for well-known lakes and reservoirs. They include the characteristics of meteorological, morphometric, climatic, trophic and hydraulic conditions but also information on dominating algal species and uses of the drainage areas. The data base contains additionally input files for the socalled reference waters, which serve to execute specific simulation runs by using the simulation system. These components of the data base permit to investigate insufficiently known lakes and reservoirs by means of analogy conclusions with the help of the expert system. The synthetical data of the data base are referred to the output files of simulation runs for specific waters and water quality problems which are stored and can be consulted.

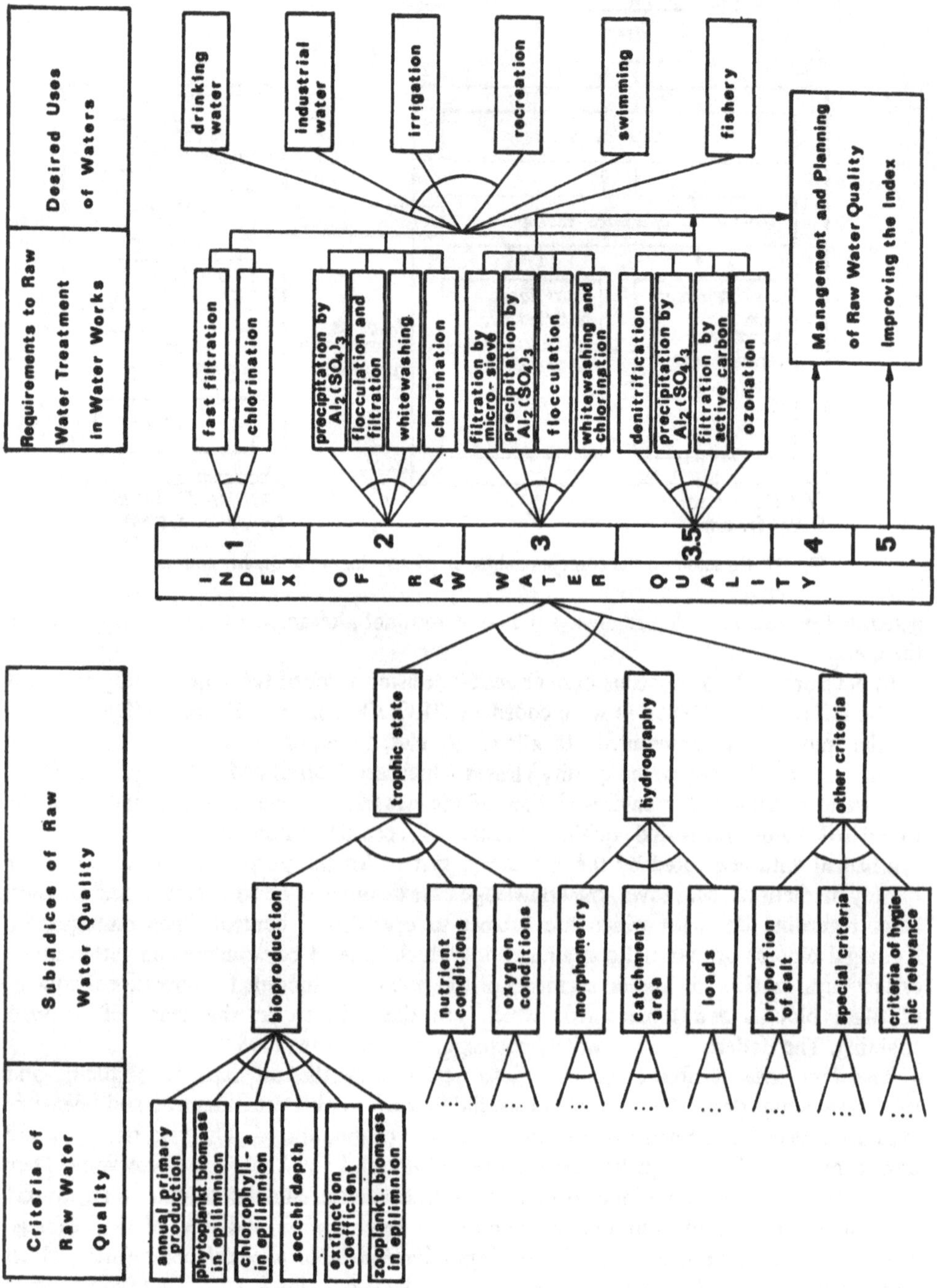

Fig. 58: AND/OR-graph of the rule-based module

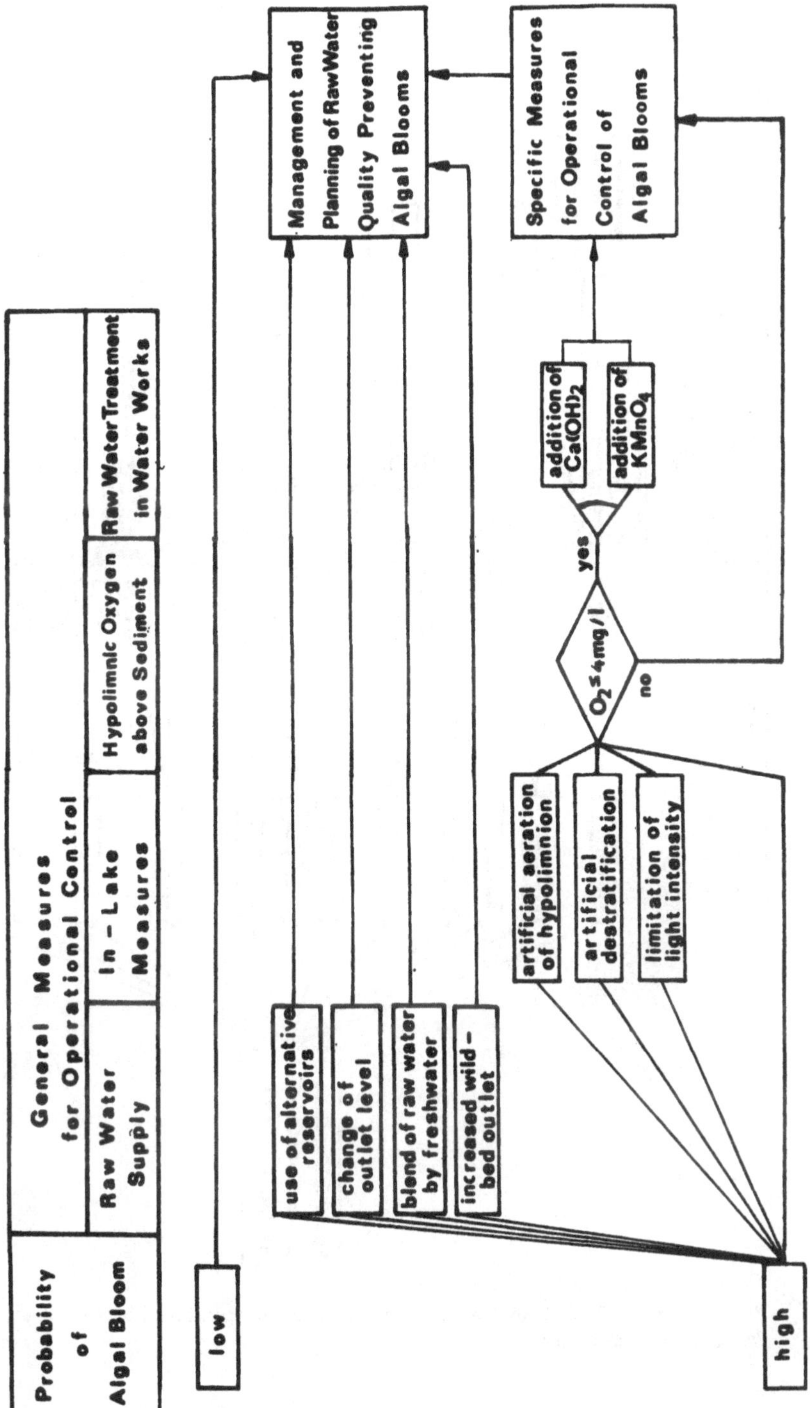

Fig. 59: AND/OR-graph of the rule-based module for operational control of algal blooms. Part 1: general measures.

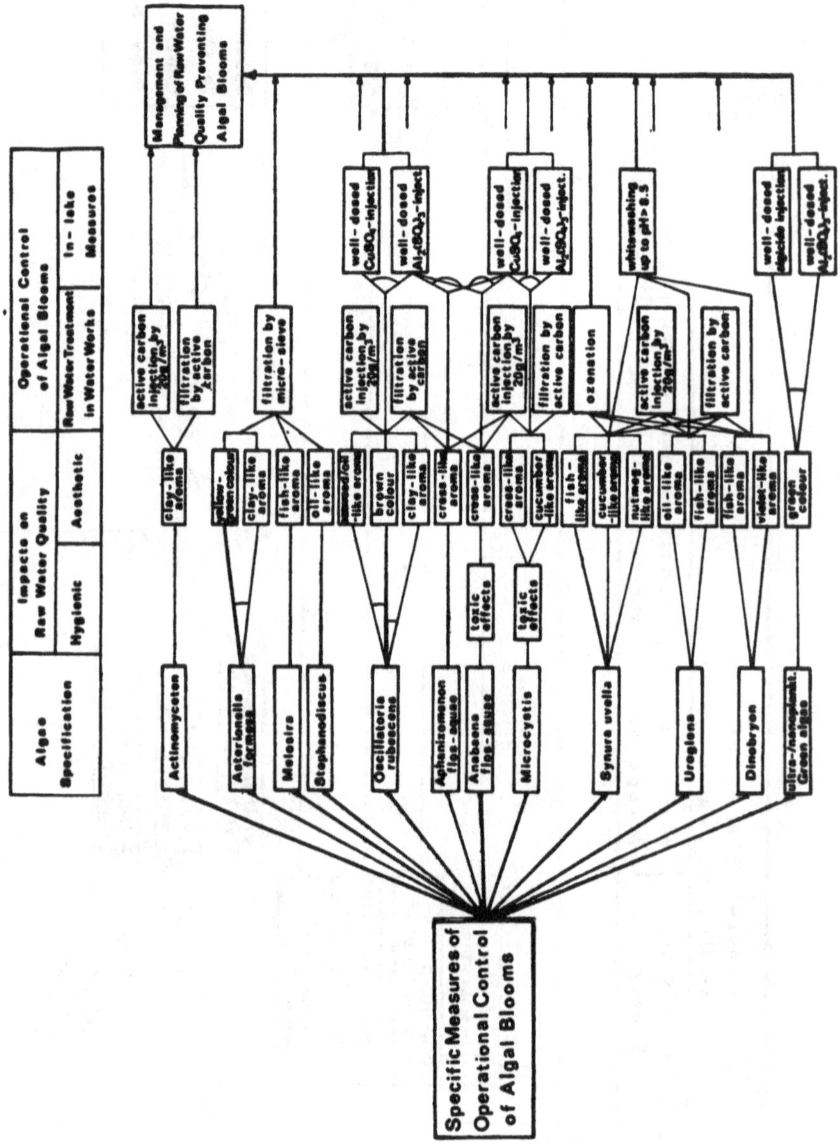

Fig. 60: AND/OR-graph of the rule-based module for operational control of algal blooms. Part 2: Specific measures.

The simulation system in Fig. 57 is based on the ecological lake model SALMO (see chapter 5), which makes available the methods of model-based decision making as introduced in chapter 6.

8.3. Conclusions

A deep expert system is developed which combines AI- and simulation methods to assist decision making in water quality control of any lakes and reservoirs at different time horizons. It places at the user's disposal the following methods in an integrated manner:

(1) Optional derivation of symptoms, consequences and control measures for undesirable events in waters like algal blooms, infections, contamination by potentially toxic including radioactive pollutants.
(2) Classification of water quality into five classes linked to potential uses of waters and recommendations for raw water treatment. The classification is based on a comprehensive standard and can be executed by measured and simulated data.
(3) Interference of judgements and decisions by analogy conclusions with the help of measured and simulated water quality data from six different and well-known reference waters.
(4) Consultation of the simulation system based on the model SALMO in order to generate predictions of consequences of control policies or environmental perturbations.

In order to qualify the expert system with respect to further decrease of uncertainty in decision making the application of fuzzy logic for knowledge representation as proposed by ZADEH (1983), BOCKLISCH et al. (1986) and SCHMUCKER (1986) is planned.

9. References

ADELSBERGER, H. H. and NEUMANN, G., 1985. Goal orientated simulation modeling using PROLOG. Proceeding of the International Workshop on Modeling and Simulation on Microcomputers, 42—47.

ANDRONOV, A. A. and PONTRJAGIN, L. S., 1937. Grubye Sistemy, Dokl. Akad. Nauk SSSR, **14**, 247—250.

BECK, M. B., 1983. Sensitivity analysis, calibration, and validation. In: ORLOB, G. T. (ed.): Mathematical Modelling of Water Quality: Streams, Lakes, and Reservoirs. J. Wiley & Sons, Chichester, New York.

BECK, M. B., 1987. Water quality modeling: a review of the analysis of uncertainty. Water Resources Research **23**, 8, 1393—1442.

BENNDORF, J., 1979a. A contribution to the phosphorus loading concept. Int. Revue ges. Hydrobiol. **64**, 2, 177—188.

BENNDORF, J., 1979b. Kausalanalyse, theoretische Synthese und Simulation des Eutrophierungsprozesses in stehenden und gestauten Gewässern. Dissertation B. Technische Universität Dresden, 1—165.

BENNDORF, J., 1980. A control mechanism of phytoplankton respiration in an eutrophic reservoir. Hydrobiologia **72**, 125—129.

BENNDORF, J., 1988. Documentation of the dynamic ecological model SALMO TU Dresden.

BENNDORF, J., KOSCHEL, R. and RECKNAGEL, F., 1985. The pelagic zone of Lake Stechlin — an approach to a theoretical model. In: CASPER, J. (ed.), Stechlin — a Natural Oligotrophic Lake. Junk Publishers, Dordrecht, 433—453.

BENNDORF, J. and RECKNAGEL, F., 1982. Problems of application of the ecological model SALMO to lakes and reservoirs having various trophic states. Ecol. Modelling **17**, 129—145.

BENNDORF, J., RECKNAGEL, F. and SCHULTZ, H., 1983. Food-web manipulation for a man-made lake (Bautzen reservoir). In: Study material for the UNEP-course Ecological Approaches to Resource Development, Land Use Management and Impact Assessment in Developing Countries, 20—41.

BENNDORF, J., UHLMANN, D. and PÜTZ, K., 1981. Strategies for water quality management in the German Democratic Republic. In: VOLLENWEIDER, R. A. (ed.), Eutrophication: A Global Problem, Part 1. Water Qual. Bull. **6**, 68—73.

BERNHARDT, H., 1981. Gewässerschutz und landwirtschaftliche Produktivität im Konflikt — die Nährstoffeliminierungsanlage des Wahnbachtalsperrenverbandes als Lösung. Wasser- und Abfallwirtschaft in Nordrhein-Westfalen, Düsseldorf, 58—69.

BERNHARDT, H. und CLASEN, J., 1981. Oligotrophication of the Wahnbach Reservoir. In: VOLLENWEIDER, R. A. (ed.), Eutrophication; A Global Problem. Part 1. Water Qual. Bull. **6**, 74—78.

BOCKLISCH, ST., ORLOVSKI, S., PESCHEL, M. and NISHIWAKI, Y. (eds.), 1986. Fuzzy Sets Applications, Methodological Approaches, and Results. Proceedings of an International Workshop on Fuzzy Sets Applications. Akademie-Verlag Berlin.

BRADLEY, R. M. and SILVA, S. A., 1977. Stabilization lagoons including experience in Brazil. Effluent Water Treat. J. **17**, 21—23, 26—29.

BREUKER, J. and WIELINGA, B., 1985. KADS: Structured knowledge acquisition for expert systems. In: Proceedings of the 5th International Workshop on Expert Systems & Applications. Avignon, Vol. 2, 887—900.

BUNGE, M., 1979. Treatise on Basic Philosophy. Volume 4. Ontology II: A World of Systems. D. Reider Publ. Comp., Dordrecht, Boston, London.

BURNS, J. R., 1977. Converting signed digraphs to Forrester schematics and converting Forrester schematics to differential equations. IEEE Transactions on Systems, Man and Cybernetics **7**, 695—707.

CANALE, R. P., DE PALMA, L. M. and VOGEL, A. H., 1976. A plankton-based food web model for Lake Michigan. In: CANALE, R. P. (ed.): Modeling Biochemical Processes in Aquatic Ecosystems, Ann. Arbor Sci., 33—74.

CASTI, J., 1979. Connectivity, Complexity and Catastrophe in Large-Scale System. J. Wiley & Sons, Chichester.

CASTI, J., 1981. Topology and crisis. Options. A IIASA News Report 3, 4—6.

CERTES, C., HUBERT, P. and LETOLLE, R., 1986. Integration of various decision tools in hydrology. Material of the UNESCO/IHP-III Symposium on Decision Support Systems and Related Methods in Water Resources Planning. Oslo, May 5—7, 1986.

CHAPRA, C. S., 1979. P-loading concept-models predicting the trophic state of waters. In: SCAVIA, D. and ROBERTSON, A. (eds.): Perspectives on Lake Ecosystem Modeling. Ann Arbor, Michigan, 66—85.

CLAPHAM, W. B., PESTEL, R. F. and ARNASZUS, H., 1979. On the scenario approach to simulation modeling for complex policy assessment and design. Policy Sciences 11, 157—177.

CLOCKSIN, W. and MELLISH, C., 1981. Programming in PROLOG. Springer-Verlag Berlin.

COLSON, G. and ZELENY, M., 1980. Multicriterion concept of risk under incomplete information. Comp. & Op. Research 7, 125—143.

CONRAD, M., 1983. Adaptability. The Significance of Variability from Molecule to Ecosystem. Plenum Press, New York, London.

DEKKER, L., 1984. Concepts for an advanced parallel simulation architecture. In: ÖREN, T. I., ZEIGLER, B. P. and M. S. ELZAS (eds.), Simulation and Model-Based Methodologies: An Integrative View. Springer-Verlag, Berlin, Heidelberg, 235—278.

DENNING, P. J., 1986. Towards a science of expert systems. IEEE Expert 1, 2, 80—83.

DITLEVSEN, O., 1981. Uncertainty Modeling. McGraw-Hill, New York.

DI TORO, D. M., OCONNOR, D. J., THOMANN, R. V. and MANCINI, J. L., 1975. Phytoplankton-zooplankton-nutrient interaction model for western Lake Erie. In: PATTEN, B. C. (ed.): Systems Analysis and Simulation in Ecology, Vol. 3, Academic Press, New York, London, 423—474.

EBENHÖH, W., 1980. A model of the dynamics of plankton patchiness. Modeling, Identification and Control 1, 2, 69—91.

EIGEN, M., 1971. Naturwissenschaften 58, 465.

EIGEN, M. and SCHUSTER, P., 1979. The Hypercycle: A Principle of Natural Self-Organization. Springer-Verlag, Berlin, Heidelberg, New York.

FEDRA, K., 1980. Mathematical modelling — a management tool for aquatic ecosystems? Helgoländer Meeresuntersuchungen 34, 221—235.

FEDRA, K., 1983. Environmental Modeling Under Uncertainty: Monte Carlo Simulation. IIASA Laxenburg, RR-83-28.

FEIGENBAUM, E. D., 1984. Knowledge engineering. The applied side of artificial intelligence. In: PAGELS, H. R. (ed.), Computer Culture. The Scientific, Intellectual and Social Impact of the Computer. The New York Academy of Sciences. New York, 91—107.

FINNEY, B. A., BOWLES, D. S. and WINDHAM, M. P., 1982. Random differential equations in water quality modeling. Water Resources Research 18, 1, 122—134.

FIERING, M. B. and HOLLING, C. S., 1974. Management and standards for perturbed ecosystems. Agro-Ecosystems 1, 301—321.

FONTAINE, T. D. and LESHT, B. M., 1987. Improving the effectiveness of environmental management decisions with optimization and uncertainty analysis techniques. In: BECK, M. B. (ed.), 1987. Systems Analysis in Water Quality Management. Pergamon Press. Oxford, New York, 31—41.

FORD, A. and GARDINER, P. C., 1979. A new measure of sensitivity for social system simulation models. IEEE Transactions on Systems, Man and Cybernetics 9, 105—114.

FORRESTER, J. W., 1961. Industrial Dynamics. MIT Press, Cambridge, Massachusetts.

FORRESTER, J. W., 1971. World Dynamics. MIT Press. Cambridge, Massachusetts.

FRANKE, D., 1984. Berechnung von Polyoptimalsteuerungen unter Verwendung hierarchischer Methoden. Diss. A. Techn. Hochschule Karl-Marx-Stadt, Fakultät für Elektroingenieurwesen, 1—112.

GDR-Standards: "Waste Water Treatment", 1982, TGL 27722/01 — "Utilization and Protection of Waters", 1982, TGL 27885/01 und 1983, TGL 27885/03. Ministry of Environmental Protection and Water Management.

GLANSDORFF, P. and I. PRIGOGINE, 1971. Thermodynamic Theory of Structure, Stability and Fluctuations. J. Wiley & Sons, New York.

GOH, B. S., 1979. The usefulness of optimal control theory to ecological problems. In: HALFON, E. (ed.): Theoretical Systems Ecology. Advances and Case Studies. Academic Press, New York.

HALFON, E. 1977. Analytical solution of the system sensitivity equation associated with a linear model. Ecol. Modelling **3**, 301—307.

HAUG, P. T., 1983. Living systems theory: conceptional basis for ecosystem modelling. In: LAUENROTH, W. K., SKOGERBOE, G. V. and FLUG, M. (eds.), 1983. Analysis of Ecological Systems: State-of-the-Art in Ecological Modelling. Elsevier, Amsterdam, Oxford, New York, 54—56.

HAYES-ROTH, B., WATERMAN, E. A. and LENAT, D. B., 1983. Building Expert Systems. Addison Wesley, Reading, Massachusetts.

HEERKLOTZ, H., 1982. Abhängigkeit der Input-Diskretisierungsintervalle des Modells SALMO von der mittleren Verweilzeit eines Standgewässers. Belegarbeit. Techn. Universität Dresden.

HOLLING, C. S., 1973. Resilience and stability of ecological systems. Annu. Rev. Ecol. Syst. **4**, 1—24.

HOLLING, C. S. (ed.). Adaptive Environmental Assessment and Management. J. Wiley & Sons, Chichester, New York.

HORN, W., 1978. Der Anteil des Crustaceenplanktons an der Verringerung des Phytoplanktongehaltes in der Saidenbachtalsperre. Diss. A, Techn. Univ. Dresden, Sektion Wasserwesen.

HOWELL, L. W., KRUTCHKOFF, R. G., PARKER, B. C., 1978. Development and use of a mathematical model for an antactic lake. Hydrobiologia **60**, 2, 129—134.

HUGHES, T. C., 1982. Lake Eutrophication Management Optimization Modeling — Approaches with Application to Lake Balaton. IIASA Laxenburg, WP-82-10.

HUSHON, J. M., 1986. Response to chemical emergencies. Environ. Sci. Technol. **20**, 2, 118—121.

HWANG, C. L., PAIDY, S. R. and YCON, K., 1980. Mathematical programming with multiple objectives: a tutorial. Comp. & Op. Research **7**, 5—31.

IKEDA, S. and ADACHI, N., 1979. A dynamic water quality model of Lake Biwa — a simulation study of the lake eutrophication. Ecol. Modelling **4**, 151—172.

IVACHNENKO, A. G., 1982. Inductive Method for the Self-Organisation of Models of Complex Systems (in Russian). Naukova Dumka, Kiew.

IVACHNENKO, A. G., KROTOV, G. I. and VISOTSKY, V. N., 1979. Identification of the mathematical model of a complex system by the self-organization method. In: HALFON, E. (ed.), Theoretical Systems Ecology. Advances on Case Studies. Academic Press, New York, San Francisco, London, 326—355.

JENTSCH, W., 1969. Digitale Simulation kontinuierlicher Systeme. R. Oldenbourg Verlag, München, Wien.

JONES, D. D., 1977. The application of catastrophe theory to ecological systems. In: INNIS, G. S. (ed.), 1977. New Corrections in the Analysis of Ecological Systems. Part 2. Simulation Council Proc. Series. La Jolla, Cal., 133—148.

JONES, D. D. and WALTERS, C. J., 1976. Catastrophe theory and fishery regulations. J. Fish. Res. Board Can. **33**, 2829—2833.

JORGENSEN, S. E. and MEJER, H., 1977. Ecological buffer capacity. Ecol. Modelling **3**, 39—61.

JOWITT, P. W. and JENKINS, W. O., 1986. The application of expert systems to river basin management. Material of the UNESCO/IHP-III Symposium on Decision Support Systems and Related Methods in Water Resources Planning. Oslo, May 5—7, 1986.

JOWITT, P. W. and J. P. LUMBERS, 1985. Expert systems to help water industry. Water Bulletin, April 26, 1985.

KALCEVA, R., OUTRATA, J., SCHINDLER, Z. and STRASKRABA, M., 1982. An optimization model for the economic control of reservoir eutrophication. Ecol. Modelling **17**, 121—128.

KALMAN, R. E., 1979. A system-theoretic critique of dynamic economic models. In: BALAKRISHNAN, A. V. and THOMA, M. (eds.), Global and Large Scale System Models, Lecture Notes in Control and Information Sciences. Vol. 19. Springer-Verlag, Berlin, Heidelberg, New York.

KAMP-NIELSEN, L., 1978. Modelling the vertical gradients in sedimentary phosphorus fractions. Verh. Internat. Verein. Limnol. **20**, 720—727.

KEMPF, J., 1980. Multiple steady states and catastrophes in ecological models. ISEM J. **2**, 55—79.

KLAPPER, H., 1977. Möglichkeiten und Effektivität der Sanierung von Talsperren und Seen. Diss. B, Techn. Univ. Dresden, Sektion Wasserwesen.

KLAPPER, H., 1983. Erarbeitung von Lösungen für die Verhinderung und Bekämpfung von Algenmassenentwicklungen in Trinkwassertalsperren und Seen. Institut für Wasserwirtschaft. Magdeburg.

KOSCHEL, R., BENNDORF, J., PROFT, G. and RECKNAGEL, F., 1983. Calcite precipitation as a natural control mechanism of eutrophication. Arch. Hydrobiol. **98**, 380—408.

KREMER, J. N. and NIXON, S. W., 1978. A Coastel Marine Ecosystem Simulation and Analysis. Springer-Verlag, Berlin, Heidelberg, New York.

KRENKEL, P. A. and NOVOTNY, V., 1980. Water Quality Management. Academic Press, New York, London.

KRUSPE, R., 1984. Aufbereitung der Eingangsdaten für das Modell SALMO zur Prognose der Entwicklung der Talsperre Eibenstock sowie kostenmäßige Erfassung wesentlicher Sanierungsmaßnahmen zur Vorbereitung einer ökonomischen Optimierung. Belegarbeit, Techn. Univ. Dresden, Sektion Wasserwesen.

KUO, F. F. and KAISER, J. F. (eds.) 1966. System Analysis by Digital Computer. J. Wiley and Sons New York, London, Sydney.

KUTAS, T. and HERODEK, S., 1982. Short description of Balaton Eutrophication Model (BEM). Working Paper MS/9, Computer and Automatic Institute, Hungarian Academy of Sciences, Budapest.

LANGE, O., 1970. Introduction to Economic Cybernetics. Pergamon Press, Oxford.

LEONOV, A., 1979. Modeling of biochemical cycles and the study of transformation processes of chemical compounds. In: JØRGENSEN, S. E. (ed.). State-of-the-Art in Ecological Modelling. Elsevier Sci. Publ. Comp., Amsterdam, Oxford, New York, 713—739.

LEVIN, S., 1979. Multiple equilibria in ecological models. In: 3rd Int. Conf. Problems in Modelling of Ecological Processes. Telavi, 1978, Moscow, Part 1, 164—230.

LEVINS, R., 1966. The strategy of model building in population biology. American Scientist **54**, 421—431.

LJAPUNOV, A. M., 1949. Problème Générale de la Stabilité du Mouvement. (1907 French translation of 1892 Russian memoires.) Reprinted as Annals of Mathematical Study, **17**. Princeton University Press, Princeton, New Jersey.

LOUCKS, D. P., STEADINGER, J. R. and HAITH, D. A., 1981. Water Resources Systems Planning and Analysis, Prentice-Hall, Englewood Cliffs, New Jersey.

LUDWIG, D., JONES, D. D. and HOLLING, C. S., 1978. Qualitative analysis of insect outbreak systems: the spruce budworm and the forest. J. Anim. Ecol. **47**, 315—332.

MAHAMAH, D. S. and BHAGAT, S. K., 1983. Use and abuse of empirical phosphorus models. In: LAUENROTH, W. K., SKOGERBOE, G. V. and FLUG, M. (eds.). Analysis of Ecological Systems: State-of-the-Art in Ecological Modelling. Elsevier, Amsterdam, Oxford, New York, 593—600.

MAJKOWSKI, J., RIDGEWAY, J. M., MILLER, D. R., 1981. Multiplicative sensitivity analysis and its role in development of simulation models. Ecol. Modelling **12**, 191—208.

MARA, D., 1979. Design verification for tropical oxidation ponds. J. Environ. Eng. Div., Am. Soc. Civ. Eng. **105**, 151—155.

MARAIS, G. R., 1966. New factors in the design, operation and performance of waste-stabilization ponds. Bull. W.H.O. **34**, 737—763.

MARSDEN, J. E. and MCCRACKEN, M., 1976. The Hopf-Bifurcation and Its Application. Springer-Verlag, New York.

MATSUMURA, T. and SAKAWA, Y., 1981. An optimization problem related to the regulation of influent nutrients in aquatic ecosystems. Int. J. Systems Sci. **12**, 5, 565—585.

MAY, R. M., 1977. Threshold and breakpoints in ecosystems with a multiplicity of stable states. Nature **269**, 471—477.

MEHRA, R. K., 1977. Catastrophe theory, nonlinear system identification and bifurcation control. In: Proc. Joint Automatic Control Conference, San Francisco, CA, 823—831.

MENSHUTKIN, V. V., KOSHOVA, O. M., ASHCHEPKOVA, L. J. and KROTOVA, V. A., 1980. A segmented model of the dynamics of the ecosystem of Lake Baikal considering three-dimensional circulation of the water. In: SWAIN, W. R. and SHANNON, V. R. (eds.). Proc. of the 2. American-Soviet Symposium on the Use of Mathematical Models to Optimize Water Quality Management. Bloomfield Hills, Mich., EPA, Duluth, Minnesota, 366—378.

MESAROVIC, M., 1979. Practical application of global modeling. In: BALAKRISHNAN, A. V. and THOMA, M. (eds.), Global and Large Scale System Models, Lecture Notes in Control and Information Sciences. Vol. 19, Berlin, Heidelberg, New York, 232 pp.

MICHAELSEN, R. and D. MICHIE, 1983. Expert systems in business. Datamation **29**, 240—246.

MILLER, J. G., 1978. Living Systems. McGraw-Hill, New York.

MONTGOMERY, R. H., LEE, D. V. and RECKHOW, K. H., 1980. A comparison of uncertainty analysis techniques: first order analysis vs. Monte Carlo simulation. IAGLR, Great Lakes, Queens College, Kingston, Ontario.

NYHOLM, N., 1978. A simulation model for phytoplankton growth and nutrient cycling in eutrophic shallow lakes. Ecol. Modelling **4**, 279—310.

OKUBO, A., 1980. Diffusion and Ecological Problems: Mathematical Models. Biomathematics, Vol. 10. Springer-Verlag, Berlin, Heidelberg, New York.

O'KEEFE, R., 1986. Simulation and expert systems — a taxonomy and some examples. Simulation **46**, 1, 10—16.

O'NEIL, R. V., GARDNER, R. H., BARNTHOUSE, L. W., SUTER, G. W., HILDEBRAND, S. G. and GEHRS, C. W., 1982. Ecosystem risk analysis: a new methodology. Environmental Toxicology and Chemistry **1**, 167—177.

ÖREN, T. I. and B. P. ZEIGLER, 1987. Artificial intelligence in modelling and simulation: Directions to explore. Simulation 48:4, 131—134.

PARK, R. A., 1974. A generalized model for simulating lake ecosystems. Simulation **23**, 2, 33—50.

PESCHEL, M., 1978. Modellbildung für Signale und Systeme. VEB Verlag Technik, Berlin.

PESCHEL, M. and MENDE, W., 1986. The Predator-Prey Model: Do We Live in a Volterra World? Akademie-Verlag Berlin.

PETZOLDT, TH., 1988. Statistische Datenanalyse zur Gültigkeits- und Risikobewertung von Modellaussagen mit SALMO. Diplomarbeit. Technische Universität Dresden, Sektion Wasserwesen, 1—83.

PIMENTEL, J. R., 1983. Real-time simulation using multiple microcomputers. Simulation **41**, 3, 93—104.

POOLE, R. W., 1979. Ecological models and the stochastic-deterministic question. In: PATIL, G. P. and ROSENZWEIG, M. (eds.). Contemporary Quantitative Ecology and Related Ecometrics. International Co-operative Publishing House. Fairland, Maryland, 251—269.

POSTON, T. and STEWART, I., 1978. Catastrophe Theory and Its Applications. Pitman, London.

PUGH, A. L., 1971. DYNAMO II User's Manual. MIT Press, Cambridge, Massachusetts.

RADTKE, E. and STRASKRABA, M., 1980. Self-optimization in a phytoplankton model. Ecol. Modelling **9**, 247—268.

RAO, A. V., 1972. Studies on algae in relation to treatment of wastewater. Doctoral thesis, Dept. of Botany, Vikram Univ., Bhopal (M. P.), India.

RECKHOW, K. H., 1979. Empirical lake models for phosphorus: development, applications, limitations and uncertainty. In: SCAVIA, D. and ROBERTSON, A. (eds.), Perspectives on Lake Ecosystem Modeling. Ann Arbor, Michigan, 193—222.

RECKHOW, K. H. and CHAPRA, ST. C., 1983. Engineering Approaches for Lake Management. Vol. 1: Data Analysis and Empirical Modeling. Butterworth Publishers, Boston, London.

RECKNAGEL, F., 1980. Systemtechnische Prozedur zur Modellierung und Simulation von Eutrophierungsprozessen in stehenden und gestauten Gewässern. Diss. A, TU Dresden Sektion Wasserwesen.

RECKNAGEL, F., 1984a. Angewandte Systemanalyse in der Wassergütebewirtschaftung von Standgewässern. Dissertation B. Technische Universität Dresden, 1—169.

RECKNAGEL, F., 1984b. A comprehensive sensitivity analysis for an ecological simulation model. Ecol. Modelling **26**, 77—96.

RECKNAGEL, F., 1985. Analysis of structural stability of aquatic ecosystems as an aid for ecosystems control. Ecol. Modelling **27**, 221—234.

RECKNAGEL, F., 1988. A deep expert system for water quality control of lake ecosystems. Statistical Journal of the United Nations Economic Commission for Europe **5**, 271—277

RECKNAGEL, F. and BENNDORF, J., 1979. An algorithmizable method assisting in the formulation of simulation models of aquatic ecosystems. ISEM. **2** (1—2), 17—32.

RECKNAGEL, F. and BENNDORF, J., 1982. Validation of the ecological simulation model SALMO. Int. Revue ges. Hydrobiol. **67**, 113—125.

RECKNAGEL, F., BENNDORF, J. and UHLMANN, D., 1986. Model-aided water quality control of standing waters. Material of the UNESCO/IHP-III Symposium on Decision Support Systems and Related Methods in Water Resources Planning. Oslo, May 5—7, 1986.

RECKNAGEL, F. and UHLMANN, D., 1986. Model-aided steady state control of waste stabilization ponds. Proc. of the International Symposium on Simulation of Systems in Biology and Medicine. Prag, Nov. 18—20, 1986.

RENGUET, E. and DUBOIS, D., 1981. A stochastic approach of the catastrophe theory. In: PAULRE (ed.), System Dynamics and Analysis of Change. North-Holland Publ. Comp., Amsterdam, 277—312.

SCAVIA, D., 1980. An ecological model of Lake Ontario. Ecol. Modelling **8**, 49—78.

SCHEIN, M., 1981. Erarbeitung ausgewählter Kopplungselemente für die kombinierte Menge-Güte-Modellierung von Speichersystemen mit einer Kombination von LBM/SALMO. Belegarbeit. Techn. Univ. Dresden, Sektion Wasserwesen.

SCHMIDT, H., 1980. Simulationstechnik. R. Oldenbourg Verlag, München, Wien.

SCHMUCKER, K. J., 1984. Fuzzy Sets, Natural Language Computations, and Risk Analysis. Computer Science Press, Rockville.

SCHULTZ, H., 1983. Untersuchungen über die Größe und Struktur des Fischbestandes sowie den Nahrungskonsum ausgewählter Fischarten in zwei Stauseen unterschiedlichen Trophiegrades. Diss. A, Techn. Univ. Dresden, Sektion Wasserwesen.

SCHWETLICK, H., SCHMIDT, C., PÖNISCH, G. und KNAUFF, L., 1977. Programmpaket Nichtlineare Gleichungen. Techn. Univ. Dresden Sektion Mathematik.

SHANNON, R. E., MAYER, R. and ADELSBERGER, H. H., 1985. Expert systems and simulation. Simulation **44**, 6, 275—284.

SHAPIRO, J., LAMARRA, V. and LYNCH, M., 1975. Biomanipulation. An ecosystem approach to lake restoration. Contribution 143 from the Limnol. Res. Centre, Univ. of Minnesota 1975.

SIX, E., 1983. Beitrag zur Erarbeitung eines hinsichtlich Wassermenge und -beschaffenheit gekoppelten Wasserbewirtschaftungsmodells mit Wahrscheinlichkeitsaussagen. Diplomarbeit, Techn. Univ. Dresden, Sektion Wasserwesen.

SMITH, J. Q., HARRISON, P. J. and ZEEMAN, E. C., 1981. Applicable catastrophe theory. III. The analysis of some discontinuous processes. In: SINHA, D. K. (ed.), Catastrophe Theory and Applications. South Asian Publishers, New Delhi/Madras, 23—52.

SOMLYODY, L., 1981. Modeling a complex environmental system: the Lake Balaton study. IIASA Laxenburg, WP-81-100.

STEELE, J. H., 1974. The Structure of Marine Ecosystems. Cambridge.

STEFIK, M., AIKINS, J., BALZER, R., BENOIT, J., BIRNBAUM, L., HAYES-ROTH, F. and SACERDOTI, E., 1982. The organization of expert systems, a tutorial. Artificial Intelligence **18**, 135—173.

STRASKRABA, M., 1979. Natural control mechanisms in models of aquatic ecosystems. Ecol. Modelling **6**, 305—321.

STRASKRABA, M., 1982. The application of predictive mathematical models of reservoir ecology and water quality. Canadian Water Resources Journal 7, 1, 283—318.

STRASKRABA, M., 1985. Simulation models as tools in ecotechnology. In: SYDOW, A., THOMA, M. and VICHNEVETSKY, R. (eds.), Proceedings of the Int. Symposium on Systems Analysis and Simulation. Akademie-Verlag, Berlin, Part 2, p. 196—199.

STRASKRABA, M. und DVORAKOVA, M., 1977. Probleme der Simulation von Talsperrenökosystemen. Acta Hydrochim. Hydrobiol. **5**, 3, 211—230.

STRAUBEL, R. und WITTMÜSS, A., 1986. Das Programmsystem REH zur rechnergestützten Entscheidungshilfe. ZKI-Information 2/86. Akademie der Wissenschaften. Berlin, 1—146.

SUDO, T., NAKASHIMA, T., AOKI, M. and KONDO, T., 1982. An adaptive array processor. IEEE Int. Solid State Conf., pp. 122—124.

SUTER, G. W., BARNTHOUSE, L. W. and O'NEILL, R. V., 1987. Treatment of risk in environmental impact assessment. Environmental Management **11**, 3, 295—303.

SWARTZMAN, G. L., 1979. A comparison of plankton simulation models emphasizing their applicability to impact assessment. Journal of Environmental Management **9**, 145—163.

SWARTZMAN, G., 1980. Evaluation of ecological simulation models. In: LEVIN, S. (ed.), Lecture Notes in Biomathematics. Vol. 33. Springer-Verlag, Berlin, Heidelberg, New York, 230—267.

THOM, R., 1975. Structural Stability and Morphogenesis. Benjamin. Reading, MA.

THOMANN, R. V., DITORO, D. M., WINFIELD, R. P. and O'CONNOR, D. J., 1975. Mathematical modeling of phytoplankton in Lake Ontario. I. Model development and verification. Grosse Ile Lab. Nat. Environ. Res. Ctr., Grosse Ile, Michigan.

THORNTON, K. W., LESSEM, A. S., FORD, D. E. and STIRGUS, C. A., 1979. Improving ecological simulation through sensitivity analysis. Simulation **32**, 155—166.

TOMOVIC, R. and VUKOBRATOVIC, M., 1972. General sensitivity theory. Modern Analytic and Computational Methods in Science and Mathematics, Vol. 35. Elsevier, Amsterdam.

UHLMANN, D., 1979a. Hydrobiology. A text for engineers and scientists. J. Wiley & Sons, Chichester, New York.

UHLMANN, D., 1979b. BOD removal rates of waste stabilization ponds as a function of loading, retention time, temperature an hydraulic flow pattern. Water Research **13**, 193—200.

UHLMANN, D., 1980. Stability and multiple steady states of hypereutrophic ecosystems. Dev. Hydrobiol. **2**, 235—247.

UHLMANN, D., 1986. Ökotechnologie — Aufgaben und Lösungen. Wissenschaft und Fortschritt **36**, 5, 108—111.

UHLMANN, D. and RECKNAGEL, F., 1982. Dimensioning of ponds for biological waste treatment. Proc. of the UNESCO/SCOPE Workshop on Ecosystems Dynamics in Freshwater Wetlands and Shallow Water Bodies, Centre of Internat. Projects GKNT Moscow, Vol. 2, 403—420.

UHLMANN, D., RECKNAGEL, F., SANDRING, G., SCHWARZ, S. and ECKELMANN, G., 1983. A new design procedure for waste stabilization ponds. Journal of the Water Pollution Control Federation **55**, 10, 1252—1255.

VINOGRADOV, M. E., MENSHUTKIN, V. V. and SHUSHKINA, E. A., 1972. On mathematical simulation of a pelagic ecosystem in tropic waters of the ocean. Marine Biology **16**, 261—268.

VOLLENWEIDER, R. A., 1976. Advances in defining critical loading levels for phosphorus in lake eutrophication. Mem. Ist. Ital. Idrobiol. **33**, 53—83.

VOLLENWEIDER, R. A., 1981. Eutrophication — a global problem. Water Quality Bulletin **6**, 3, 59—62.

VOLLENWEIDER, R. and DILLON, P., 1974. The application of the phosphorus loading concept to eutrophication research. National Research Council Canada. Associate Committee on Scientific Criteria for Environmental Quality. Report No. 7.

VOLTERRA, V., 1926. Variazione e fluttuazini del numero d'individui in specie animali conviventi. Mem. Accad. Nazionale Lincei (ser. **6**) 2, 31—113.

WALTERS, C. J., PARK, R. A. and KOONCE, J. F., 1980. Dynamic models of lake ecosystems. In: LE CREN, E. D. and McDONELL, R. H. (eds.), The Functioning of Freshwater Ecosystems. IBP 22. Cambridge University Press, 455—479.

WASHBURN, R. B. and MEHRA, R. K., 1980. Bifurcation analysis and dynamic stability. Joint Automatic Control Conference, San Francisco, California, TP3-A.

WATANABE, M., HARLEMAN, D. R. F. and VASILIEV, O. F., 1983. Two- and three-dimensional mathematical models for lakes and reservoirs. In: ORLOB, G. T. (ed.), Mathematical Modeling of Water Quality: Streams, Lakes and Reservoirs. J. Wiley & Sons, Chichester, New York, 274—336.

WIERZBICKI, A. P. and KURCYUSZ, S., 1977. Projection on a cone, penalty functionale and duality theory for problems with inequality constraints in Hilbert space. SIAM J. Contr. Optimiz. **15**, 25—56.

WILLIAMS, B. K., 1982. Optimal stochastic control in natural resource management: framework and examples. Ecol. Modelling **16**, 275—297.

WILSON, A. G., 1981. Catastrophe Theory and Bifurcations. Applications to Urban and Regional Systems. Croom Helm London, Univ. of California Press, Berkeley, Los Angeles.

WINSTON, P. and HORN, B., 1981. LISP. Addison-Wesley. Reading, Massachusetts.

DE WIT, C. T. and RABBINGE, R., 1979. Systems analysis and dynamic simulation. EPPO Bull. **9**, 3, 149—153.

ZADEH, L. A., 1979. A theory of approximate reasoning. In: HAYES, J. E., MICHIE, D. and MIKULICH, L. I. (eds.), Machine Intelligence 9. J. Wiley & Sons, New York.

ZADEH, L. A., 1983. The rule of fuzzy logic in the management of uncertainty in expert systems. Fuzzy Sets and Systems **11**, 199—227.

ZEEMAN, E. C., 1976. Catastrophe theory. Sci. Am. **243**, 65—83.

ZEEMAN, E. C., 1977. Catastrophe Theory. Selected Papers, 1972—1977. Wesley, MA.

ZEEMAN, E. C., 1981. Applicable catastrophe theory 1. Stability of ships. In: SINHA, D. K. (ed.), Catastrophe Theory and Applications. South Asian Publishers, New Delhi/Madras, 11—19.

ZEIGLER, B. P., 1976. Theory of Modelling and Simulation. J. Wiley & Sons, New York, Sidney, Toronto.

Fachbereichsstandard „Abwasserbehandlung", 1982. Natürlich belüftete Abwasserteichanlagen, Anwendung und Bemessung. TGL 28722/01, Ministerium für Umweltschutz und Wasserwirtschaft, Berlin, 1—6.

Fachbereichsstandard „Nutzung und Schutz der Gewässer", Stehende Binnengewässer, 1982. TGL 27885/01, Ministerium für Umweltschutz und Wasserwirtschaft, Berlin, 1—16.

Fachbereichsstandard „Nutzung und Schutz der Gewässer", 1982. Wassergütebewirtschaftung Seen. TGL 27885/03, Ministerium für Umweltschutz und Wasserwirtschaft, Berlin, 1—36.

10. Index

Zeitfracht Medien GmbH
Ferdinand-Jühlke-Straße 7
99095 Erfurt, Deutschland
produktsicherheit@kolibri360.de